Azure Security Cookbook

Practical recipes for securing Azure resources and operations

Steve Miles

BIRMINGHAM—MUMBAI

Azure Security Cookbook

Associate Group Product Manager: Mohd Riyan Khan
Senior Editor: Divya Vijayan
Technical Editor: Nithik Cheruvakodan
Copy Editor: Safis Editing
Project Coordinator: Ashwin Kharwa
Proofreader: Safis Editing
Indexer: Rekha Nair
Production Designer: Joshua Misquitta
Marketing Coordinator: Marylou De Mello

First published: March 2023
Production reference: 1230223

Published by Packt Publishing Ltd.
Livery Place
35 Livery Street
Birmingham
B3 2PB, UK.

ISBN 978-1-80461-796-0
www.packtpub.com

This book is my contribution to the worldwide technical learning community, and I would like to thank all of you who are investing your valuable time in learning new skills and committing to reading this book.

Contributors

About the author

Steve Miles, aka *SMiles* or *Mr. Analogy*, is a Microsoft Azure MVP, MCT, multi-cloud, and hybrid technologies author and technical reviewer with over 20 years of experience in security, networking, data center infrastructure, managed hosting, and cloud solutions. His experience comes from working in end user, reseller channel, and vendor spaces, with global networks, data and app security vendors, global telco hosters, and colocation and data center services providers, as well as in managed hosting and hardware distribution.

His roles have included network security architect, global solutions architect, public cloud security solutions architect, and Azure practice technical lead. He currently works for a leading multi-cloud distributor based in the UK and Dublin in a cloud and hybrid technology leadership role.

Most happy in front of a whiteboard, he prefers to speak using illustrations. He is renowned for his analogies for breaking down complex technologies and concepts into everyday, real-world scenarios.

His first Microsoft certification was on Windows NT. He is an MCP, MCITP, MCSA, and MCSE for Windows Server and many other Microsoft products. He also holds multiple Microsoft Fundamentals, Associate, Expert, and Specialty certifications in Azure Security, Identity, Network, M365, and D365. He also holds multiple security and networking vendor certifications, including PRINCE2 and ITIL, and is associated with industry bodies such as the CIF, ISCA, and IISP. Finally, as part of the multi-cloud aspect, he has experience with GCP and AWS, is Alibaba-Cloud-certified, and is a nominated Alibaba Cloud MVP.

About the reviewer

Peter De Tender has over 25 years of experience in architecting and deploying Microsoft solutions, starting with Windows NT4/Exchange 5.5 in 1996. Since early 2012, he started shifting to cloud technologies and quickly embraced Azure, working as a cloud architect and trainer. In September 2019, he joined Microsoft's prestigious Microsoft Technical Trainer team, providing Azure readiness workshops to its top customers and partners across the globe. Recently having relocated to Redmond, WA, he continues in this role. Given his past Azure MVP and community passion, he is still actively involved in public speaking, technical writing, and mentoring/coaching.

You can follow Peter on Twitter at `pdtit` and check out his technical blog at `007ffflearning.com`.

Thanks to Steve, for trusting in my love for Azure. Also, a big thanks to my wife for supporting me in realizing my dreams.

Patrick Lownds is a master-level solution architect working for Pointnext Advisory & Professional Services, in the Hybrid IT COE, for Hewlett Packard Enterprise (HPE), and is based in London, UK. He currently works with the most recent versions of Windows Server and System Center and has participated in the Windows Server, System Center, and Microsoft Azure Stack Early Adoption Program. He is a community blogger for HPE and tweets in his spare time. He can be found on Twitter at patricklownds.

Table of Contents

2

Securing Azure Networks   49

3

Securing Remote Access   123

4

Securing Virtual Machines   163

5

Securing Azure SQL Databases   189

6

Securing Azure Storage   211

Part 2: Azure Security Tools

7

Using Advisor   241

8

Using Microsoft Defender for Cloud   255

9

Using Microsoft Sentinel   287

10

Using Traffic Analytics   331

Preface

With the increase in usage of cloud platforms and with many companies embracing a hybrid workforce, new threat vectors are emerging and cyber-attacks are increasing.

A new security model mindset is required more than ever, a model that thinks beyond traditional device-based and network-perimeter-based security. We need to adopt a holistic approach to security, starting with insights and highlighting identity as the new control and security pane.

This book is a recipe-based guide to help you become well versed with Azure security features and tools.

You will start with learning important Azure security features such as identities, virtual machines, networks, storage, databases, and remote access. Then, you will dive into Defender for Cloud, Microsoft Sentinel, and other related tools to safeguard your identities, infrastructure, apps, and data.

Every chapter is independent, takes up important problems, and provides solutions, including those related to implementing and operating security features and tools.

By the end of the book, you will have learned to secure Azure cloud platform resources and have a guide you can use to solve specific day-to-day challenges.

Who is this book for

This book targets security-focused professionals looking to protect Azure resources using the native Azure platform security features and tools.

A solid understanding of the fundamental security concepts and prior exposure to Azure will help you understand the key concepts covered in the book more effectively.

This book also benefits those aiming to take the Microsoft certification exam with a security element or focus.

What this book covers

Chapter 1, *Securing Azure AD Identities*, teaches users how to secure and protect Azure AD identities. We will break down the chapter into sections on reviewing Azure AD identity secure scores, implementing Identity and Access Management on Azure AD tenants, implementing Azure AD Password Protection, implementing Self-Service Password Reset, implementing the Azure AD security defaults, implementing Azure AD Multi-Factor Authentication, implementing Conditional Access policies, implementing Azure AD Identity Protection, and implementing Azure AD Privileged Identity Management.

Chapter 2, Securing Azure Networks, explains how to secure and protect Azure networks. We will break down the chapter into sections covering implementing Network Security Groups, implementing Azure Firewall, implementing Azure Web Application Firewall, and implementing Azure DDoS.

Chapter 3, Securing Remote Access, focuses on how to secure and protect remote access. We will break down the chapter into sections covering implementing the Azure Bastion service, implementing Azure Network Adapter, and implementing **Just-in-Time (JIT)** VM access.

Chapter 4, Securing Virtual Machines, takes securing and protecting Azure VMs as its subject. We will break down the chapter into sections on implementing VM Update Management, implementing VM Microsoft antimalware, and implementing Disk Encryption for Azure VMs.

Chapter 5, Securing Azure SQL Databases, discusses how to secure and protect Azure databases. We will break down the chapter into sections on implementing a service-level IP firewall, implementing a private endpoint, and implementing Azure AD authentication and authorization.

Chapter 6, Securing Azure Storage, breaks down how to secure and protect Azure storage. We will break down the chapter into sections covering implementing security settings on storage accounts, implementing network security, and implementing encryption.

Chapter 7, Using Advisor, explores how to secure and protect Azure environments using the Advisor recommendations engine. We will break down the chapter into sections on the security recommendations and secure scores and perform the implementation of recommendations.

Chapter 8, Using Microsoft Defender for Cloud, demonstrates the components of Defender for Cloud, as well as how to enable the enhanced security features of Defender for Cloud, add a regulatory standard to the regulatory compliance dashboard, and assess environment regulatory compliance against the added standard.

Chapter 9, Using *Microsoft Sentinel*, walks through enabling Microsoft Sentinel and how to review the components, create automation, set up a data connector, and create an analytics rule.

Chapter 10, Using Traffic Analytics, covers the implementation of Traffic Analytics.

To get the most out of this book

For this book, the following are required:

- *A device with a browser, such as Edge or Chrome, to access the Azure portal at* `https://portal.azure.com`

- *An Azure AD tenancy and Azure subscription; you can use an existing one or sign up for free:* `https://azure.microsoft.com/en-us/free`

- *A Global Admin role for the Azure AD tenant*

- *An Owner role for the Azure subscription*

Download the color images

We also provide a PDF file that has color images of the screenshots and diagrams used in this book. You can download it here: `https://packt.link/fPcIW`.

Conventions used

There are a number of text conventions used throughout this book.

`Code in text`: Indicates code words in text, database table names, folder names, filenames, file extensions, pathnames, dummy URLs, user input, and Twitter handles. Here is an example: A device with a browser, such as Edge or Chrome, to access the Azure portal: `https://portal.azure.com`

Any command-line input or output is written as follows:

```
Get-AzVmDiskEncryptionStatus
```

Bold: Indicates a new term, an important word, or words that you see onscreen. For instance, words in menus or dialog boxes appear in **bold**. Here is an example: "We will start by looking at **Active Directory (AD)**."

> **Tips or important notes**
> Appear like this.

Get in touch

Feedback from our readers is always welcome.

General feedback: If you have questions about any aspect of this book, email us at `customercare@packtpub.com` and mention the book title in the subject of your message.

Errata: Although we have taken every care to ensure the accuracy of our content, mistakes do happen. If you have found a mistake in this book, we would be grateful if you would report this to us. Please visit `www.packtpub.com/support/errata` and fill in the form.

Piracy: If you come across any illegal copies of our works in any form on the internet, we would be grateful if you would provide us with the location address or website name. Please contact us at `copyright@packt.com` with a link to the material.

If you are interested in becoming an author: If there is a topic that you have expertise in and you are interested in either writing or contributing to a book, please visit `authors.packtpub.com`

Share your thoughts

Once you've read *Azure Security Cookbook*, we'd love to hear your thoughts! Scan the QR code below to go straight to the Amazon review page for this book and share your feedback.

https://packt.link/r/1804617962

Your review is important to us and the tech community and will help us make sure we're delivering excellent quality content.

Download a free PDF copy of this book

Thanks for purchasing this book!

Do you like to read on the go but are unable to carry your print books everywhere?

Is your eBook purchase not compatible with the device of your choice?

Don't worry, now with every Packt book you get a DRM-free PDF version of that book at no cost.

Read anywhere, any place, on any device. Search, copy, and paste code from your favorite technical books directly into your application.

The perks don't stop there, you can get exclusive access to discounts, newsletters, and great free content in your inbox daily

Follow these simple steps to get the benefits:

1. Scan the QR code or visit the link below

https://packt.link/free-ebook/9781804617960

2. Submit your proof of purchase

3. That's it! We'll send your free PDF and other benefits to your email directly

Part 1: Azure Security Features

In this part, we will go through recipes that provide complete coverage of the skills and knowledge required to implement and operate native Azure platform security features.

This part includes the following chapters:

- *Chapter 1, Securing Azure AD Identities*
- *Chapter 2, Securing Azure Networks*
- *Chapter 3, Securing Remote Access*
- *Chapter 4, Securing Virtual Machines*
- *Chapter 5, Securing Azure SQL Databases*
- *Chapter 6, Securing Azure Storage*

Securing Azure AD Identities

Azure Active Directory (**Azure AD**) is a multi-tenant cloud-based identity and access management solution that is part of Microsoft's **Entra Identity platform** product family.

You can read more about *Entra* and its integrated *hybrid* and *multi-cloud identity* and *access solutions* family at the following Microsoft site: `https://www.microsoft.com/en-us/security/business/microsoft-entra`.

In this chapter, you will learn how to *secure* and *protect* Azure AD identities.

We will break down this chapter into sections that cover how you can review your environments, including security posture, tenant-level identity and access management, password management and protection, security defaults, multi-factor authentication, and Conditional Access. We will then look at implementing Identity Protection and *Identity Management* services.

By the end of this chapter, you will have covered the following recipes to create secure Azure AD identities:

- Reviewing Azure AD Identity Secure Score
- Implementing Azure AD tenant Identity and Access Management
- Implementing Azure AD Password Protection
- Implementing Self-Service Password Reset
- Implementing Azure AD security defaults
- Implementing Azure AD multi-factor authentication
- Implementing Conditional Access policies
- Implementing Azure AD Identity Protection
- Implementing Azure AD Privileged Identity Management

Introduction to Azure Identity Services

Before we look at any recipes, we will first introduce some concepts surrounding *Microsoft Identity services*. This will assist us in establishing a foundation of knowledge to build upon. We will start by looking at **Active Directory** (**AD**).

What is AD?

AD provides **Identity and Access Management** (**IAM**) and **Information Protection** services for traditional Windows Server environments. It was first included with *Windows Server 2000* as an installable service.

AD provides different services in its portfolio and is used as a generic and *umbrella term* in many cases.

These individual services in Azure AD include the following:

- **AD Domain Services** (**AD DS**)
- **AD Federation Services** (**AD FS**)
- **AD Certificate Services**
- **AD Rights Management Services**

In this next section, we will introduce Azure AD and look at its relationship with AD, a similar name but with different functions, capabilities, and use cases.

When is AD not AD? When it is Azure AD!

Before we go any further, we should clear one thing up: there is a common misconception that **Azure AD** must just be a cloud-based **Software-as-a-Service** (**SaaS***) version*, but it is **not**!

It is easy enough why people (*wrongly)* think this may be the case; after all, **Exchange Online** and **SharePoint Online** are indeed exactly that, *SaaS versions* of their traditional infrastructure deployed platforms; if only it were that simple, though.

In many ways, **Azure AD** is like **AD** on the surface; they are both **Identity Providers** (**IDPs**) and provide **IAM** controls. Still, at the same time, they function differently and don't yet provide a complete parity of capabilities, although quite close.

It is worth noting that Azure AD is constantly evolving to meet the requirements and demands of authentication and authorization of workloads and services to bring capabilities in line with those available in AD, such as **Kerberos realms** within Azure AD.

At the time of publishing this book, you *cannot use* Azure AD to 100% replace the provided capabilities of AD.

Depending on the scenario, it may be the case that your environments will never be 100% cloud-based for identity services. You may remain with *Hybrid identity services* – that is, both AD and Azure AD coexist in a connected and synchronized state.

What is Azure AD?

Azure AD is a *SaaS identity management solution* that is *fully managed* and provides functions such as an *IDP* and *IAM* for managing and securing access to resources based on **Role-Based Access Control (RBAC)**.

As Azure AD is provided as a *fully managed service*, there is no installable component such as **Windows Servers and Domain Controllers (DC)**; *zero infrastructure* needs to be deployed by you.

The primary cloud authentication protocol used by Azure AD is based around using **OpenID**, **OAuth**, and **Graph**, whereas AD uses **Kerberos** and **NTLM**.

What is Hybrid Identity?

The hybrid identity approach allows you to *synchronize objects*, such as *user objects* and their *passwords*, between AD and Azure AD *directories*.

The main driver for hybrid identity within an organization is legacy AD-integrated applications that do not support cloud identity authentication protocols.

This capability provides users access to *AD authenticated*, and *Azure AD authenticated* using a single **Common Identity** and password.

The password synced to Azure AD is a *hash* of the stored *hashed password*; passwords are never stored in Azure AD, only the password hash. This capability is referred to as **same sign-on**, meaning you will be prompted each time to enter the *same* credentials when you wish to authenticate to resources.

This capability should not be confused with **single sign-on (SSO)**, which *does not* prompt you again when accessing resources. The following diagram shows the relationship between AD and Azure AD:

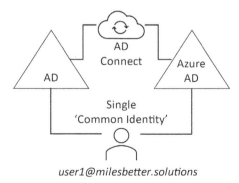

Figure 1.1 – AD and Azure as a relationship

Azure AD Connect is a free downloadable tool that *synchronizes* objects between AD and Azure AD's IDP directories; this establishes *hybrid identities*. Azure AD Connect provides additional functionality and capabilities and allows for **Self-Service Password Reset (SSPR)** through additional configuration.

You can continue learning more, should you wish, about hybrid identities and Azure AD Connect, by going to `https://learn.microsoft.com/en-us/azure/active-directory/hybrid/whatis-azure-ad-connect`.

Technical requirements

For this chapter, the following are required for the recipes:

- A machine with a modern browser such as Edge or Chrome and internet access; this machine can be a client or server operating system. We will use a Windows 10 Microsoft Surface laptop with a Chrome browser for the recipe examples.

- An Azure AD tenancy; you may use an existing one or sign up for free: `https://azure.microsoft.com/en-us/free`.

- Access to the *Global Administrator* role for the tenancy.

- Some cloud-only test user created accounts as part of the Azure AD tenancy.

- You will require Azure AD Premium licenses or trial licenses. The following steps will guide you on activating a free trial if you do not already have a license:

  I.   From the Azure portal, go to **Azure AD | Licenses | All products**, then click **Try/Buy** from the *top toolbar*.

  II.   Select the **AZURE AD PREMIUM P2** free trial and click **Activate**:

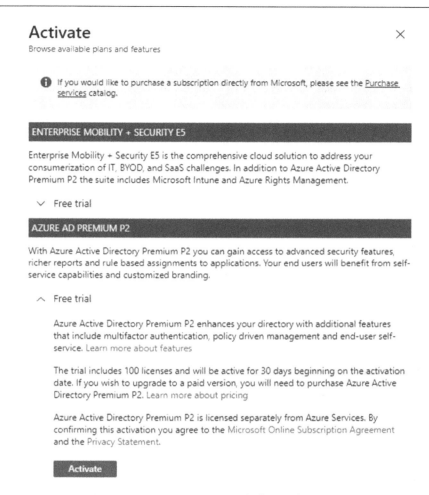

Figure 1.2 – Azure AD Premium P2 free trial activation

Reviewing Azure AD Identity Secure Score

Azure AD Identity Secure Score enables you to make informed decision-making to protect your Azure AD tenancy.

This recipe will teach you how to monitor and *improve* your Azure AD Identity Secure Score.

We will take you through reviewing the Azure AD Identity Secure Score dashboard for your Azure AD tenancy environments and look at the actionable insights available to improve your secure score and security posture.

Getting ready

This recipe requires the following:

- A device with a browser, such as Edge or Chrome, to access the Azure portal: `https://portal.azure.com`
- You should sign into the Azure portal with an account with the **Global Administrator** role

How to do it...

This recipe consists of the following tasks:

- Reviewing Identity Secure Score
- Updating the improvement actions status

Task – Reviewing Identity Secure Score

Perform the following steps:

1. From the Azure portal, go to **Azure Active Directory | Security | Identity Secure Score**.

   Alternatively, in the search bar, type `azure ad identity secure score`; click on **Azure AD Identity Secure Score** from the list of services shown.

2. You will now see the **Identity Secure Score** blade.

3. The top section of the **Identity Secure Score** screen represents your *identity security posture*:

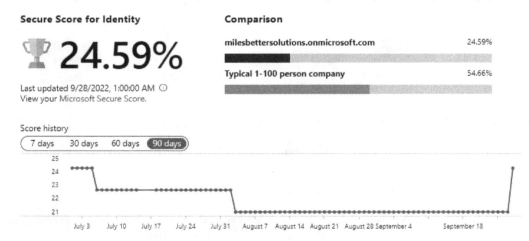

Figure 1.3 – Secure Score screen

This area of the screen shows three aspects to review:

- **Secure Score for Identity** is a percentage of your alignment with Microsoft's best practice security recommendations

- **Comparison** is your security posture management compared to other tenants of a similar size

- **Score history** is a trend graph over time

4. The lower section of the **Identity Secure Score** screen provides a list of recommended and possible security **Improvement actions**.

    Each recommended improvement action has a **Score Impact**, **User Impact**, **Implementation Cost**, **Max Score** possible, and **Current Score**:

Improvement actions

↓ Download   ≡≡ Columns

Name ↑↓	Score Impact ↑	Current Score ↑↓	Max Score ↑↓	User Impact ↑↓	Implementation C... ↑↓	Status ↑↓
Require multifactor authentication for administrative roles	16.39%	0	10	Low	Low	To address
Ensure all users can complete multifactor authentication	14.75%	0	9	High	High	To address
Enable policy to block legacy authentication	13.11%	0	8	Moderate	Moderate	To address
Do not expire passwords	13.11%	8	8	Moderate	Low	Completed
Protect all users with a user risk policy	11.48%	0	7	Moderate	Moderate	To address
Protect all users with a sign-in risk policy	11.48%	0	7	Moderate	Moderate	To address
Enable password hash sync if hybrid	8.20%	5	5	Low	Low	Completed
Do not allow users to grant consent to unreliable applications	6.56%	0	4	Moderate	Low	To address
Use least privileged administrative roles	1.64%	1	1	Low	Low	Completed
Designate more than one global admin	1.64%	1	1	Low	Low	Completed
Enable self-service password reset	1.64%	0	1	Moderate	Moderate	To address

Figure 1.4 – The Improvement actions screen

5. Click **Download**; you can access the improvement actions in a CSV file:

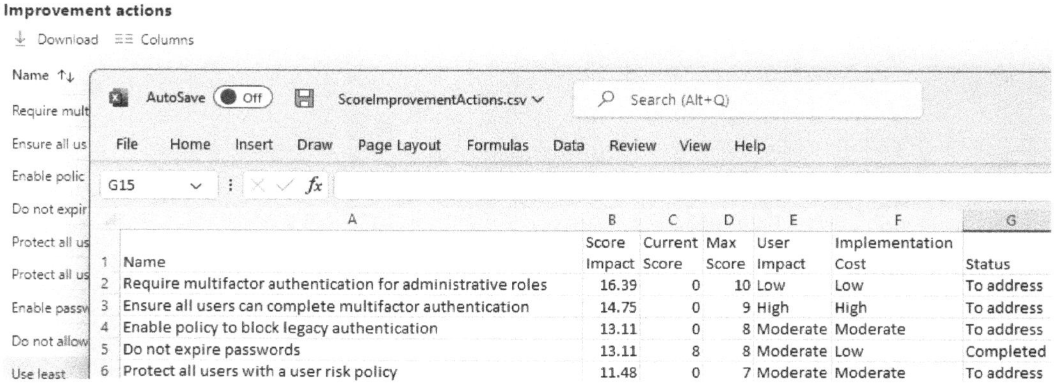

Improvement actions

↓ Download   ≡≡ Columns

	A	B	C	D	E	F	G
1	Name	Score Impact	Current Score	Max Score	User Impact	Implementation Cost	Status
2	Require multifactor authentication for administrative roles	16.39	0	10	Low	Low	To address
3	Ensure all users can complete multifactor authentication	14.75	0	9	High	High	To address
4	Enable policy to block legacy authentication	13.11	0	8	Moderate	Moderate	To address
5	Do not expire passwords	13.11	8	8	Moderate	Low	Completed
6	Protect all users with a user risk policy	11.48	0	7	Moderate	Moderate	To address

Figure 1.5 – Improvement actions download

6. By clicking on an **Improvement action**, you can see further information:

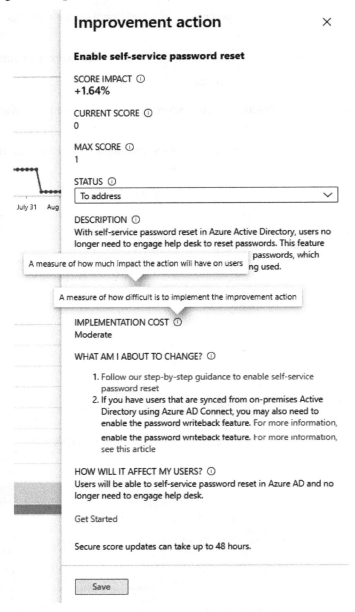

Figure 1.6 – Improvement actions information

With that, you have reviewed Identity Secure Score. In the next task, we will update the status of improvement actions.

Task – Updating the improvement actions status

Perform the following steps:

1. Select an **Improvement action** and click to open it.

2. From the **Improvement action** screen, on the **STATUS** section, select the status you wish to update the action to and then click **Save**:

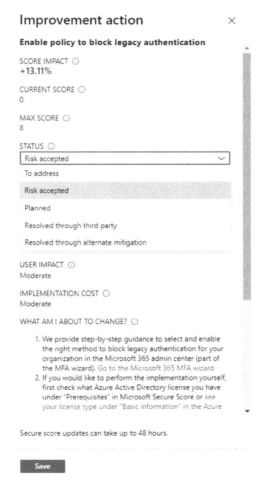

Figure 1.7 – Improvement actions status options

With that, you have updated the status of improvement actions. This concludes the hands-on tasks for this recipe.

How it works...

In this recipe, we reviewed the information presented in the Azure AD identities Secure Score and took action from available insights.

- The Azure ID Identity Secure Score overlaps with the identity score used for the *Microsoft secure score*, which means the recommendations will be the same.

- The Azure AD Identity Secure Score provides a value of between **1%** and **100%**, representing how well your Azure AD tenancy is secured based on Microsoft's best practices and *recommendations*.

You can also see actionable improvement insights on how your score can be improved and each improvement's impact on the secure score.

The dashboard and a score history timeline show a comparison of your environment's Azure AD tenancy to a tenancy of the same size and industry average.

Your environment's Azure AD tenancy identity settings are compared with best practice recommendations once a day (approx 1:00 A.M. PST); changes made to an improvement action *may not* be reflected in the score for up to **48 hours**.

See also

Should you require further information, you can refer to the following Microsoft Learn articles:

- What is the identity secure score in Azure Active Directory?: `https://learn.microsoft.com/en-us/azure/active-directory/fundamentals/identity-secure-score`

- Azure Active Directory fundamentals documentation: `https://learn.microsoft.com/en-us/azure/active-directory/fundamentals`

Implementing Azure AD tenant Identity and Access Management

Account compromise is one of the biggest threat vectors to protect against, and those with privileged access roles will be the focus of attacks. There are often too many users assigned privileged accounts, with more access than is required for a user to carry out their role. There is often insufficient RBAC in place, and the principle of least privilege should be adopted for these privileged administrator roles.

While we need to limit the number of user accounts that have the Global Administrator role, there should also not be a single point of compromise for the Global Administrator role. Having more than one account with the Global Administrator role is important. It is crucial to have an emergency account in case of a breach or conditional access lockout of a Global Administrator role assigned. Global Administrator role accounts can use a buddy system to monitor each other's accounts for signs of a breach.

This recipe will teach you to ensure you only have the users assigned with the least privileges required for their role and ensure you have a minimum of two accounts assigned the Global Administrator role.

We will take you through the steps to implement these tasks.

Getting ready

This recipe requires the following:

- A device with a browser, such as Edge or Chrome, to access the Azure portal: `https://portal.azure.com`
- You should sign in with an account that has the **Global Administrator** role

How to do it...

This recipe consists of the following tasks:

- Implementing least privileged administrative roles
- Designating more than one Global Administrator

Task – implementing least privileged administrative roles

Perform the following steps:

1. From the Azure portal, go to **Azure Active Directory | Roles and administrators**.

2. From the **All roles** section, select the **Global administrator** role:

Figure 1.8 – Azure AD Roles and Administrators screen

3. From the **Assignments** section, identify only the accounts required to have the **Global Administrators** role; ensure you have at *least two* or no more than *five* accounts with the **Global Administrator** role.

   Select a user for users who no longer require the **Global Administrator** role and then click **Remove assignments** from the top toolbar:

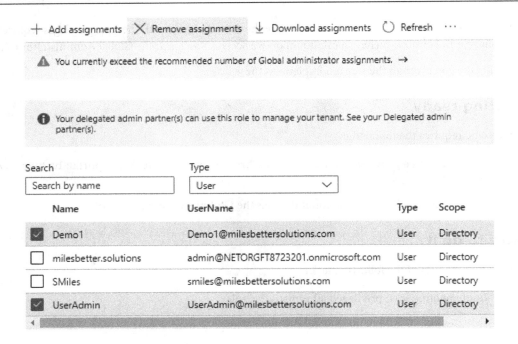

Figure 1.9 – The Remove assignments screen

4.   From **Azure Active Directory | Roles and administrators | All roles | Global administrator**, we can now see that the user has been removed from the **Global Administrator** role:

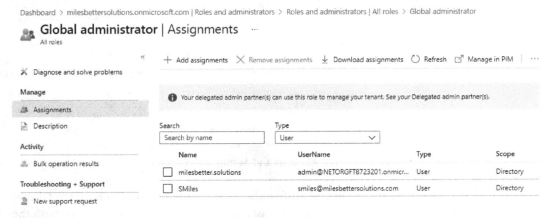

Figure 1.10 – Global Administrator Assignments screen

5.   To reassign least privileged admin users to roles required to complete their tasks, navigate to **Azure Active Directory | Users**. Select and click the *users* to assign *roles*.

6. From the **User** blade for the user selected to assign a directory role, go to **Assigned roles** from the **Manage** section and click **Add assignments**:

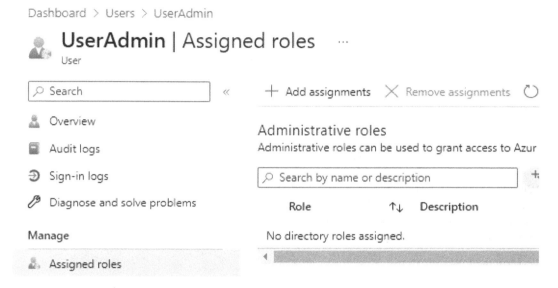

Figure 1.11 – The Assigned roles screen

7. From the **Directory roles** pop-up screen, locate the *directory role* you wish to assign from the list of all available roles; select the *directory role* to assign and click **Add**:

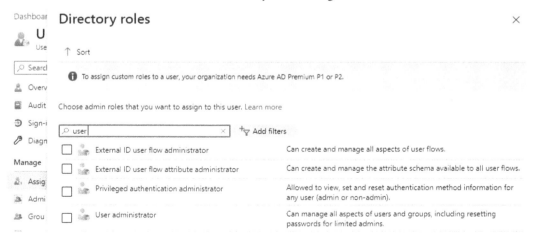

Figure 1.12 – The Directory roles assignment screen

8. Your user will now have the required *least privileged admin* role assigned and no longer have the highly privileged **Global Administrator** role:

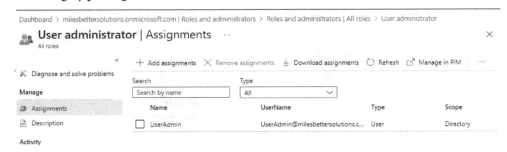

Figure 1.13 – User administrator | Assignments

With that, you have learned how to use least privileged roles. In the next task, we will designate more than one Global Administrator for the tenancy.

Task – designating more than one Global Administrator

Perform the following steps:

1. From the Azure portal, go to **Azure Active Directory | Roles and administrators | All roles | Global Administrator**.

2. From the **Assignments** blade, click **Add assignments** and locate the user(s) to add to the **Global Administrators** role:

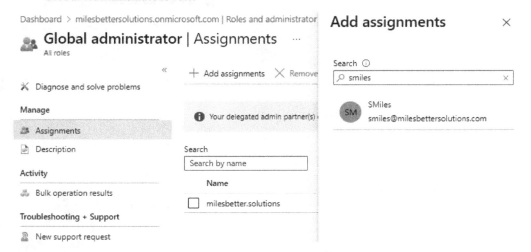

Figure 1.14 – Global administrator – the Add assignments screen

3. Select the user, and then click **Add**:

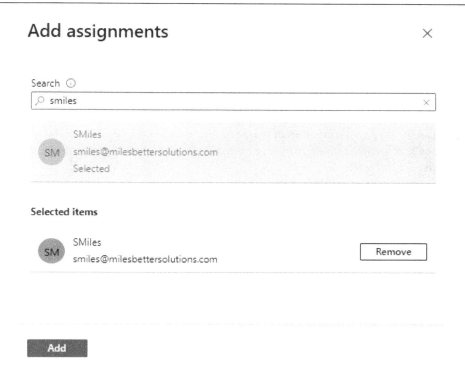

Figure 1.15 – Global administrator – The Add assignments screen

4. You will now see that the user(s) have been assigned the **Global Administrator** role:

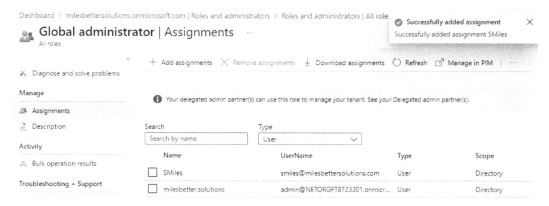

Figure 1.16 – Global administrator | Assignments

With that, you have created more than one Global Administrator role. This concludes the hands-on tasks for this recipe.

How it works...

In this recipe, we looked at limiting the number of users with the Global Administrator role and ensuring you only had the users assigned with the least required privileges for their role. In our example, we removed the Global Administrator role from a user and reassigned them to the User Administrator role, which was the least privileges required for their tasks.

We then ensured you had a minimum of two accounts assigned the Global Administrator role by adding a user to this role. The Microsoft recommendation is for a minimum of two users and no more than five for this role.

There's more...

Azure AD user accounts with the *highest privileged* role of **Global Administrator** will be the primary goal for compromise by bad actors. This is because this role has access to every administrative setting in your environment's Azure AD tenancy at the **read** and **modify** permission level.

Microsoft recommends that you assign user accounts with *less privileged* roles. This limits the user's scope of permissions through **RBAC** to only be able to do what a user needs to do for their job function.

The following are some of the many roles that can be considered to reduce the use of the *Global Administrator* role but still have enough access for a user to be able to perform their duties:

- Application Administrator
- Authentication Administrator
- Azure DevOps Administrator
- Azure Information Protection Administrator
- Billing Administrator
- Compliance Administrator
- Conditional Access Administrator
- Directory Readers
- Exchange Administrator
- SharePoint Administrator
- Privileged Role Administrator
- Security Administrator
- User Administrator

Should you require further information on least privileged roles, you can refer to the following Microsoft Learn articles:

- Assigning Azure roles using the Azure portal: `https://learn.microsoft.com/en-us/azure/role-based-access-control/role-assignments-portal`
- Azure AD built-in roles: `https://learn.microsoft.com/en-us/azure/active-directory/roles/permissions-reference`
- What are the default user permissions in Azure Active Directory?: `https://learn.microsoft.com/en-us/azure/active-directory/fundamentals/users-default-permissions`
- Least privileged roles by task in Azure Active Directory: `https://learn.microsoft.com/en-us/azure/active-directory/roles/delegate-by-task`

See also

Should you require further information, you can refer to the following Microsoft Learn articles:

- Best practices for Azure AD roles: `https://learn.microsoft.com/en-gb/azure/active-directory/roles/best-practices`
- Restrict member users' default permissions: `https://learn.microsoft.com/en-gb/azure/active-directory/fundamentals/users-default-permissions#restrict-member-users-default-permissions`
- Azure Active Directory fundamentals documentation: `https://learn.microsoft.com/en-us/azure/active-directory/fundamentals`

Implementing Azure AD Password Protection

Users often make poor choices when creating passwords, making them easy targets and victims of dictionary-based attacks.

This recipe will teach you how to implement Azure AD password protection in your environment's AD tenancy. We will take you through customizing your smart lockout threshold and creating a global and custom banned password list.

Getting ready

This recipe requires the following:

- A device with a browser, such as Edge or Chrome, to access the Azure portal: `https://portal.azure.com`
- You should sign in with an account that has the **Global Administrator** role
- We will use Azure AD Premium licenses for this and future recipes

How to do it...

This recipe consists of the following task:

- Configuring password protection

Task – configuring password protection

Perform the following steps:

1. From the Azure portal, go to **Azure Active Directory** and then click **Security** under the **Manage** section from the *side menu*.

2. Select **Authentication Methods** under the **Manage** section from the *side menu*.

3. Select **Password protection** under the **Manage** section from the *side menu*.

4. From the **Custom smart lockout** section, set the **Lockout threshold** and **Lockout duration in seconds** properties as required; review the information in the tooltips by clicking on the **i** symbol:

Figure 1.17 – Azure AD Premium P2 free trial activation

5. From the **Custom banned password** section, select **Yes**, enter strings that are to be banned, and click **Save**; review the information in the tooltips by clicking on the **i** symbol. It can take several hours to apply the band password list:

When enabled, the words in the list below are used in the banned password system to prevent easy-to-guess passwords.

Enforce custom list ⓘ   [**Yes**   No]

Custom banned password list ⓘ

ils

events

Figure 1.18 – Azure AD Premium P2 free trial activation

With that, you have configured password protection. This concludes the hands-on tasks for this recipe.

How it works...

You only need to add key terms such as **password** or **contoso** and the algorithm will automatically consider and block all variants of common character substitutions, such as **Pa$sw0rd1!** or **C@ntos0!**.

The banned password list may have a maximum of 1,000 key terms. The minimum length of a term string is 4 characters, where 16 characters is the maximum and are case-sensitive.

This recipe looked at customizing your smart lockout threshold to protect against brute-force attack methods. We also looked at creating a global and custom banned password list to protect against dictionary and password spray attacks and enforce the use of strong passwords.

Both of these measures, when implemented, can offer significant protection for your environment's Azure AD tenancy.

See also

Should you require further information, you can refer to the following Microsoft Learn articles:

- Eliminate bad passwords using Azure Active Directory Password Protection: `https://learn.microsoft.com/en-us/azure/active-directory/authentication/concept-password-ban-bad`

- Azure Active Directory fundamentals documentation: `https://learn.microsoft.com/en-us/azure/active-directory/fundamentals`

Implementing a Self-Service Password Reset

Users will sometimes forget their passwords; to prevent intervention by an Azure AD administrator, a **self-service password reset (SSPR)** can be implemented. This allows users to click on the **Can't access your account?** link on the sign-in page for the portal or Microsoft Cloud service they are trying to access.

This recipe will teach you how to implement SSPR in your environment's AD tenancy. We will take you through enabling SSPR for a selected scope and review the available settings, then carry out a user registration for SSPR and test its operation to confirm the function is working.

Getting ready

This recipe requires the following:

- A device with a browser, such as Edge or Chrome, to access the Azure portal: `https://portal.azure.com`
- You should sign in with an account that has the **Global Administrator** role
- Optionally, pre-create an **Azure AD Security group** called **SSPR-Test-Group** and add members to test with

How to do it...

This recipe consists of the following task:

- Configuring Self-Service Password Reset

Task – configuring Self-Service Password Reset

Perform the following steps:

1. From the Azure portal, go to **Azure Active Directory** and then click **Password** under the **Manage** section from the *side menu*.
2. From **Properties**, under the **Manage** section from the *side menu*, choose **Selected** under **Self-service password reset enabled**; review the information in the tooltips on this page by clicking on the **i** symbol:

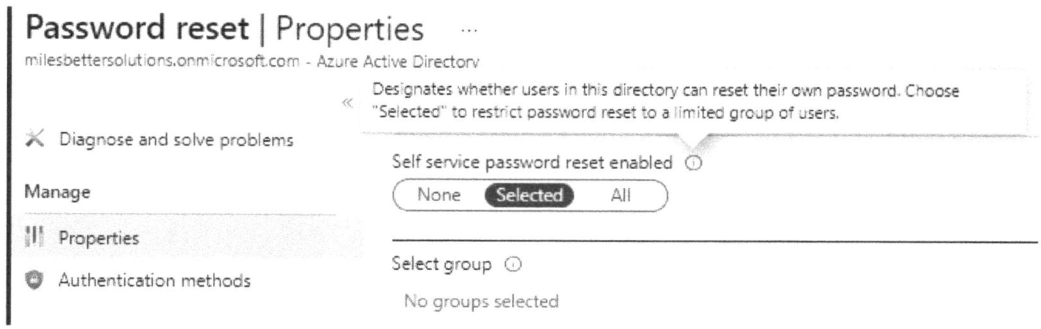

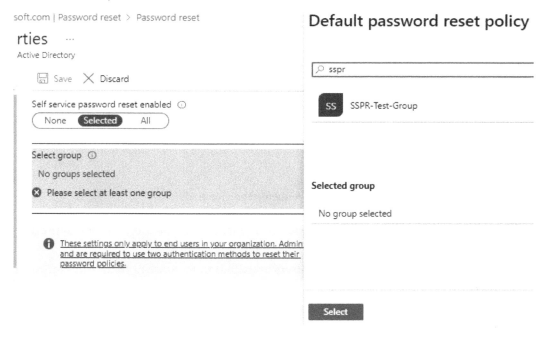

Figure 1.19 – Password reset | Properties

3. Click on the **No groups Selected** hyperlink and then browse and select the group to enable SSPR. Then, click **Save**:

Figure 1.20 – Password reset selected groups

4. From **Authentication methods**, under the **Manage** section from the *side menu*, select as required the **Number of methods required to reset** setting.

5. Then, select as required the **Methods available to users** setting:

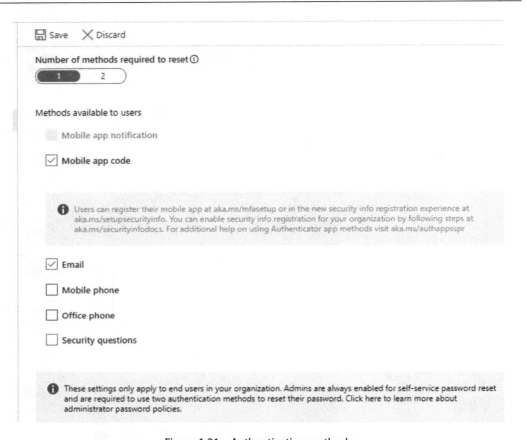

Figure 1.21 – Authentication methods

6. From **Registration**, under the **Manage** section from the *side menu*, select **Yes** for **Require users to register when signing in?**.

7. Select the **Number of days before users are asked to re-confirm their authentication information** setting as required.

8. From **Notifications**, under the **Manage** section from the *side menu*, select **Notify users on password resets?** as required.

9. From **Notifications**, under the **Manage** section from the *side menu*, select the **Notify users on password resets?** and **Notify all admins when other admins reset their password?** settings as required.

10. From **Customization**, under the **Manage** section from the *side menu*, select the **Customize helpdesk link?** and **Custom helpdesk email or URL** settings as required.

11. Review the *settings configured* from **Administrator Policy** in the **Manage** section from the *side menu*.

With that, you have configured SSPR. This concludes the hands-on tasks for this recipe.

How it works...

In this recipe, we looked at how we can implement SSPR when users forget their password for a portal or Microsoft Cloud service they are trying to access.

This prevents intervention from an Azure AD administrator, which reduces the burden on these roles and also protects against loss of productivity.

See also

Should you require further information, you can refer to the following Microsoft Learn articles:

- Tutorial: Enable users to unlock their accounts or reset passwords using Azure Active Directory SSPR: `https://learn.microsoft.com/en-us/azure/active-directory/authentication/tutorial-enable-sspr`

- Azure Active Directory fundamentals documentation: `https://learn.microsoft.com/en-us/azure/active-directory/fundamentals`

Implementing Azure AD security defaults

The perimeter vanishes with the rise in hybrid working and a remote workforce on unsecured devices outside of secure corporate networks. Now, it is commonplace to be targeted by identity-related attacks such as password spray and phishing. However, with basic security adoption, such as blocking legacy authentication and **multi-factor authentication** (**MFA**), 99.9% of these identity-related attacks can be stopped. However, we must balance security with productivity.

Because security can require skills and money, Microsoft is providing no-cost preconfigured secure settings by default to provide a basic level of security for everybody.

This recipe will teach you how to implement the Azure AD security defaults in your environment's AD tenancy.

Getting ready

This recipe requires the following:

- A device with a browser, such as Edge or Chrome, to access the Azure portal: `https://portal.azure.com`

- You should sign into the Azure portal with an account with the **Global Administrator, Security Administrator, or Conditional Access Administrator** role

How to do it....

This recipe consists of the following task:

- Enabling security defaults

Task – enabling security defaults

Perform the following steps:

1. From the Azure portal, go to **Azure Active Directory** and click **Properties** in the **Manage** section from the *side menu*.

2. Then, click the **Manage Security Defaults** hyperlink, select **Yes** under **Enable security defaults**, and click **Save**:

Figure 1.22 – The Enable security defaults screen

With that, you have enabled security defaults. This concludes the hands-on tasks for this recipe.

How it works...

In this recipe, we looked at enabling security defaults in your environment's Azure AD tenancy.

The security defaults are Microsoft-recommended security mechanisms with preconfigured security settings that, once enabled, are automatically enforced in your tenant to protect against the most common identity-based attacks.

The following are the enforced settings:

- Azure MFA for all users and administrators
- Blocking of legacy authentication protocols
- Protection of privileged access activities, such as Azure portal access

See also

Should you require further information, you can refer to the following Microsoft Learn articles:

- Security defaults in Azure AD: `https://learn.microsoft.com/en-us/azure/active-directory/fundamentals/concept-fundamentals-security-defaults`
- Azure Active Directory fundamentals documentation: `https://learn.microsoft.com/en-us/azure/active-directory/fundamentals`

Implementing Azure AD multi-factor authentication

We must adopt a **zero-trust** strategy in the perimeter-less world of cloud services and hybrid working more than ever. This means that we must **assume breach** and **never trust, always verify**.

Azure AD MFA provides an additional layer of defense; we never trust a single authentication method and must assume that the traditional password method has been compromised. Microsoft studies show that when you implement MFA, your accounts are more than 99.9% less likely to be compromised.

This recipe will teach you how to implement Azure AD MFA in your environment's AD tenancy.

Getting ready

This recipe requires the following:

- A device with a browser, such as Edge or Chrome, to access the Azure portal: `https://portal.azure.com`.
- You should sign into the Azure portal with an account with the **Global Administrator** role.

- You will require Azure AD Premium licenses or trial licenses.

- If you have Security Defaults enabled, you will automatically have MFA enabled for all users and administrators using the free benefits of Azure AD. Using one of the paid Azure AD Premium licenses provides additional capabilities, such as the additional authentication methods of verification codes, text messages, or phone calls, as well as the following:

  - **Azure AD Premium P1**: This license includes **Azure Conditional Access** for MFA

  - **Azure AD Premium P2**: This license adds **risk-based Conditional access** to MFA through **Information Protection**

How to do it...

This recipe consists of the following task:

- Configuring MFA

Task – configuring MFA

Perform the following steps:

1. From the Azure portal, go to **Azure Active Directory**, click **Security** in the **Manage** section from the *side menu*, and then click **Multifactor authentication**.

2. From the **Multifactor authentication | Getting started** blade, click the **Additional cloud-based multifactor authentication settings** hyperlink under the **Configure** section heading:

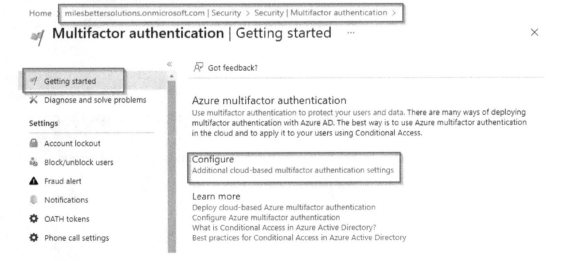

Figure 1.23 – Multifactor authentication | Getting started

3. Two tabs are available from the new **multi-factor authentication** page that opens; select the user's tab and then **users** to enable MFA:

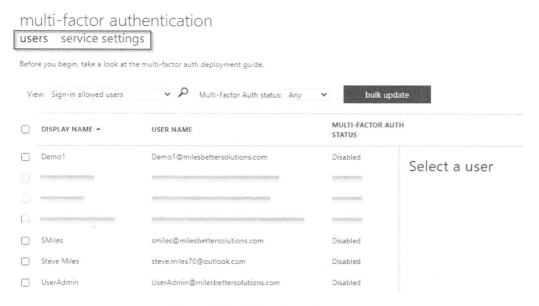

Figure 1.24 – MFA configuration screen

4. From the **user** pane on the right, click on the **Manage user settings** hyperlink in the **quick steps** section:

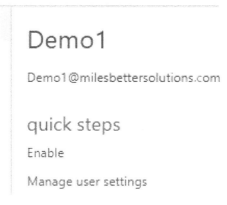

Figure 1.25 – MFA selected user pane

5. On the **Manage user settings** pop-up screen, select any of the three options as required and then select **save**:

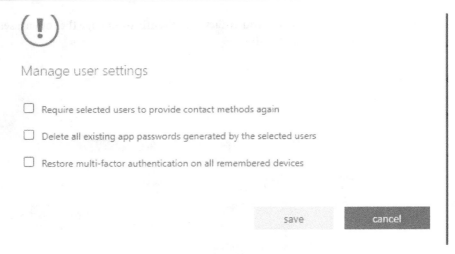

Figure 1.26 – Manage user settings pop-up screen

6. Click **Enable** on the user pane screen from *Step 4* of this recipe. From the **About enabling multi-factor auth** pop-up screen that appears, read the provided links, click **enable multi-factor auth**, and click **close** on the **Updates successful** screen.

7. To disable a user for MFA, select the user from the **user** pane, click **Disable** in the **quick steps** section, select **Yes** on the pop-up screen, and click **Close**:

Figure 1.27 – Disabling MFA for a user

8. You may bulk update enabling users for MFA by selecting the **bulk update** button and uploading a CSV file; a template file will be provided that you can download.

9. Once the **user** tab configuration is complete, select the **service settings** tab in the **multi-factor authentication** browser window:

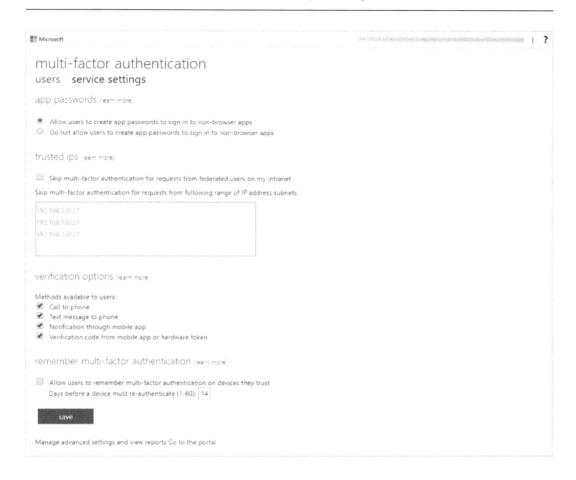

Figure 1.28 – The service settings tab's settings

10. From the **service settings** screen, set the required options and click **save**. Note the **verification options** section.

With that, you have configured MFA. This concludes the hands-on tasks for this recipe.

How it works...

In this recipe, we looked at how to enable Azure AD MFA in our environment's Azure AD tenancy to provide an additional layer of security for users to sign to protect their identity from compromise.

Azure AD MFA requires us to provide one or more additional factors as a method to authenticate in addition to the password factor.

We can use the following authentication factors:

- Something we know (*password)*
- Something we own (*device)*
- Something we are (*biometrics)*

See also

Should you require further information, you can refer to the following Microsoft Learn articles:

- Secure Azure Active Directory users with Multi-Factor Authentication: `https://learn.microsoft.com/en-us/training/modules/secure-aad-users-with-mfa/`
- Features and licenses for Azure AD Multi-Factor Authentication: `https://learn.microsoft.com/en-us/azure/active-directory/authentication/concept-mfa-licensing`
- Azure Active Directory fundamentals documentation: `https://learn.microsoft.com/en-us/azure/active-directory/fundamentals`

Implementing Conditional Access policies

There must be a balance of protecting an organization's resources while ensuring every user, wherever they are, is empowered to be productive whenever.

To further strengthen our Azure AD identities, we can use insights from identity-driven signal data to make informed access control decisions and then use those decisions to enforce access policies.

MFA works alongside Conditional Access to provide further granular control of access.

Conditional Access is based on an IF/THEN approach. This approach means that IF signal information collected from the sign-in process matches certain criteria, THEN decisions are made based on the information as to whether access will be *allowed* or *blocked*.

Conditional Access will also determine whether the user will be required to perform additional authentication methods or take other actions, such as resetting their password. This is represented in the following diagram:

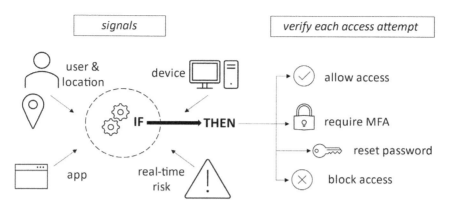

Figure 1.29 – Conditional Access concept

The following are some common Conditional Access policies:

- Require MFA for all users

- Require MFA for Microsoft portals/services access

- Require password reset for risky users

- Block the use of legacy authentication protocols

- Require hybrid-joined or compliant devices

- Allow or deny from specific locations

This recipe will teach you how to implement Conditional Access policies in your environment's AD tenancy. We will take you through enabling conditional access policies and configuring them to restrict user access to apps based on if a set of conditions have been met.

Getting ready

This recipe requires the following:

- A device with a browser, such as Edge or Chrome, to access the Azure portal: `https://portal.azure.com`.

- You should sign into the Azure portal with an account with the **Global Administrator** role.

- You will require Azure AD Premium licenses or trial licenses.

- If you have Security Defaults enabled, you will automatically have MFA enabled for all users and administrators using the free benefits of Azure AD. Using one of the paid Azure AD Premium licenses provides additional capabilities such as the additional authentication methods of verification codes, text messages, or phone calls, as well as the following:

  - **Azure AD Premium P1**: This license includes Azure Conditional Access for MFA

  - **Azure AD Premium P2**: This license adds risk-based Conditional access to MFA

How to do it...

This recipe consists of the following task:

- Configuring Conditional Access

Task – configuring Conditional Access

Perform the following steps:

1. From the Azure portal, go to **Azure Active Directory**, click **Security** in the **Manage** section from the *side menu*, and then click **Conditional Access** in the **Protect** section.

2. Click **+ New Policy** from the *top toolbar* in the **Conditional Access Policies** blade:

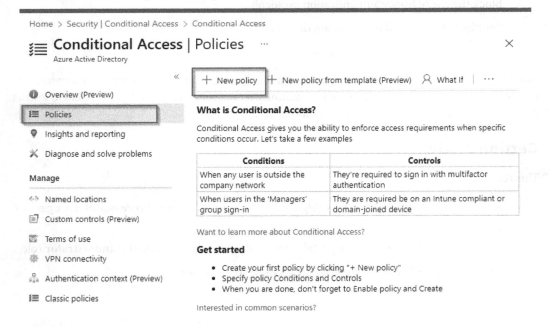

Figure 1.30 – Conditional Access | Policies

3. Select a **Name** for your policy from the **New conditional access policy** blade.

4. From the **Assignments** section, select which *users and groups* this policy will apply to:

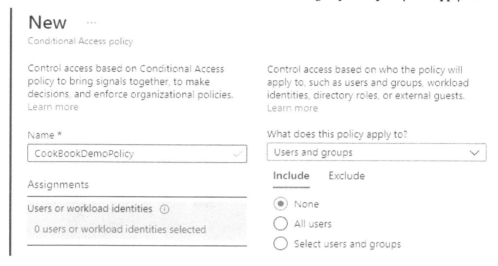

Figure 1.31 – User settings

5. From the **Cloud apps or actions** section, select whether this policy will apply to **Cloud apps** or **Actions**; we will select **Cloud apps**:

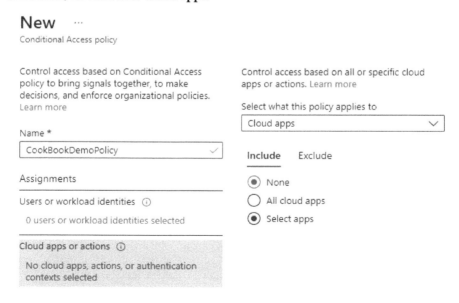

Figure 1.32 – Apps setting

6. From the **Include** tab, we will click **Select apps**, search for **Azure Management**, tick the check box next to **Microsoft Azure Management app** in the list, and click **Select**. Note the warning dialog box about not locking yourself out:

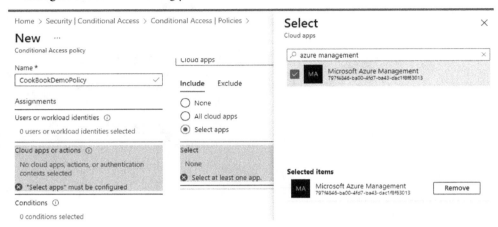

Figure 1.33 – App selection

7. Click the **Conditions** settings, set any required conditions, or leave it unconfigured:

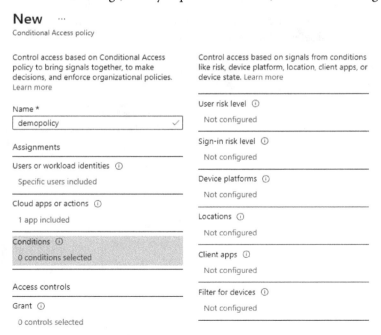

Figure 1.34 – Conditions settings

8. From **Grant**, under the **Access controls** section, click on **0 controls selected**, set it to **Grant access**, tick **Require multifactor authentication**, and then click **Select**:

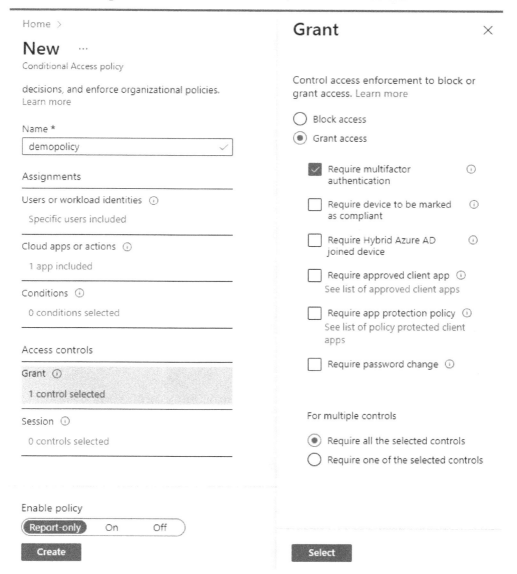

Figure 1.35 – Access settings

9. In the **Enable policy** section, leave it set to **Report-only**, then click **Create**.

10. Your policy will now appear in the policies list:

Figure 1.36 – Access policies list

With that, you have configured Conditional Access. This concludes the hands-on tasks for this recipe.

How it works...

In this recipe, we looked at how we can implement Conditional Access policies in addition to MFA to layer on an additional layer of defense while maintaining the users' productivity needs.

We configured a Conditional Access policy to a set of selected users (or groups) that required MFA when they accessed the Azure portal; this was enabled by selecting the Microsoft Azure Management app.

See also

Should you require further information, you can refer to the following Microsoft Learn articles:

- What is Conditional Access?: `https://learn.microsoft.com/en-us/azure/active-directory/conditional-access/overview`

- Azure AD Conditional Access documentation: `https://learn.microsoft.com/en-us/azure/active-directory/conditional-access`

- Conditional Access: Cloud apps, actions, and authentication context: `https://learn.microsoft.com/en-us/azure/active-directory/conditional-access/concept-conditional-access-cloud-apps`

- Azure Active Directory fundamentals documentation: `https://learn.microsoft.com/en-us/azure/active-directory/fundamentals`

Implementing the Azure AD Identity Protection service

We need solutions that provide remediation actions based on threat intelligence insights. Using policies, we can detect and respond to identity-based threats automatically; this allows us to react quicker and does not rely on human operator intervention.

This recipe will teach you how to implement Azure AD Identity Protection in your environment's AD tenancy.

We will take you through setting up risk policies, MFA registration policies, investigation, reports, and how to remediate identified risks.

Getting ready

This recipe requires the following:

- A device with a browser, such as Edge or Chrome, to access the Azure portal: `https://portal.azure.com`
- You should sign in to the Azure portal with an account with the **Global Administrator** role
- You will require Azure AD Premium licenses or trial licenses

How to do it...

This recipe consists of the following task:

- Configuring Identity Protection

Task – configuring Identity Protection

Perform the following steps:

1. From the Azure portal, go to **Azure Active Directory**, click **Security** in the **Manage** section from the *side menu*, and then click **Identity Protection** in the **Protect** section.
2. From the **Identity Protection** blade, click **User risk policy**:

Figure 1.37 – User risk policy

3. From **Assignments**, click **All users**, review the available options, and select as required. You can set it to **include** or **exclude**.

4. From **User risk**, select the **risk level controls** options to be enforced: **High**, **Medium and above**, or **Low and above**. Then, click **Done**.

5. Click **Block access** from the **Access** section under **Controls** and select the controls to be enforced. You can set it to **Block** or **Allow** access and **Require password change**. Then, click **Done**:

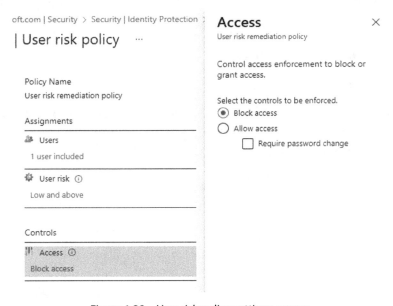

Figure 1.38 – User risk policy settings screen

6. Select **On** under **Enforce policy**, and then click **Save**.

7. Complete the same steps but this time for **Sign-in risk policy**:

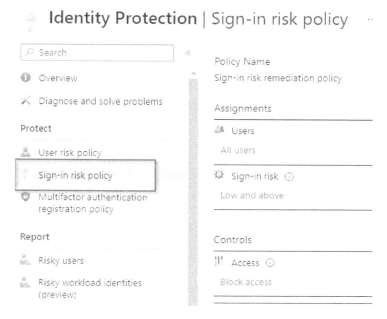

Figure 1.39 – Sign-in risk policy settings screen

With that, you have configured Identity Protection. This concludes the hands-on tasks for this recipe.

How it works...

This recipe looked at how to implement Azure AD Identity Protection.

A risk policy will monitor for identity risks, which, when detected, enforce remediation measures, which are the controls that have been set, such as blocking or allowing access and requiring a password change by the user.

See also

Should you require further information, you can refer to the following Microsoft Learn articles:

- Manage Azure AD Identity Protection: `https://learn.microsoft.com/en-us/training/modules/manage-azure-active-directory-identity-protection`

- Azure Active Directory fundamentals documentation: `https://learn.microsoft.com/en-us/azure/active-directory/fundamentals`

Implementing Azure AD Privileged Identity Management

To protect your environment's Azure AD tenancy and improve your security posture, you should implement a robust privileged identity protection strategy for roles and resources.

This recipe will teach you to implement Azure AD **Privileged Identity Management** (**PIM**) in your environment's AD tenancy.

We will take you through configuring a user to be assigned a privileged access role in your Azure AD tenancy so that the user's activity may be controlled.

Getting ready

This recipe requires the following:

- A device with a browser, such as Edge or Chrome, to access the Azure portal: `https://portal.azure.com`
- You should sign into the Azure portal with an account with the **Global Administrator** role
- You will require Azure AD Premium licenses or trial licenses

How to do it...

This recipe consists of the following task:

- Configuring Privileged Identity Management

Task – configuring Privileged Identity Management

Perform the following steps:

1. From the Azure portal, search for **Azure AD Privileged Identity Management** and select **access**.

2. From **Azure AD Privileged Identity Management**, select **Azure Resources** and click **Discover resources**:

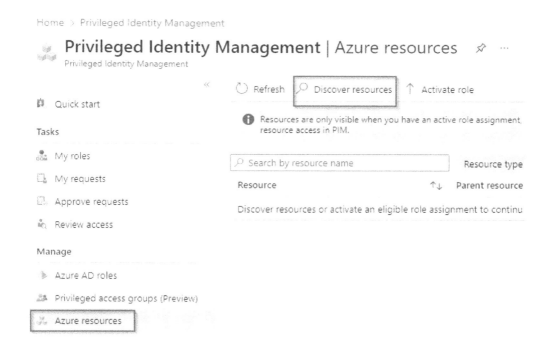

Figure 1.40 – The Privileged Identity Management screen

3.   Select your **Subscription** from the **Azure resources** blade and click **Manage resource** from the *top toolbar*. Click **OK** on the pop-up screen, then *close* the **Discovery** page:

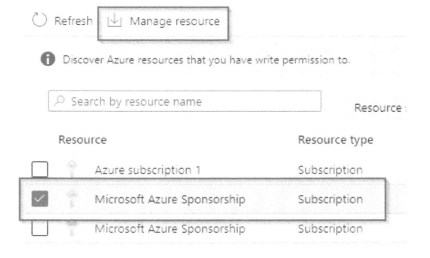

Figure 1.41 – The Azure resources blade

4. Click the subscription listed on the **Azure resources** page; the **Overview** page will open. From the *left menu*, click **Roles** in the **Manage** section:

Figure 1.42 – Manage resources screen

5. From the **Roles** blade, click **+ Add assignments** from the *top toolbar*.

6. From the **Select role** drop-down menu, select a *role* you want to be controlled via PIM. In our example, we will select the **Azure Arc Kubernetes Admin** role:

Figure 1.43 – Select role

7. Click the **No member selected under Select member(s)** hyperlink and search and select a user from your *Azure AD tenant* to be assigned this role:

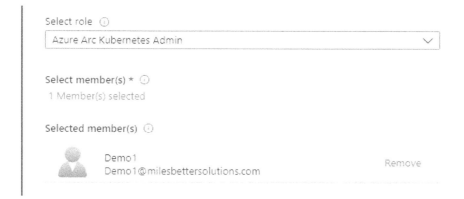

Figure 1.44 – Select member(s)*

8. Click **Next >**.

9. Select **eligible** under **assignment type** from the **setting** tab and set the **assignment start** and **end date/times** properties. Then, click **Assign**.

10. You will now see information from the **Overview** page regarding this new assignment:

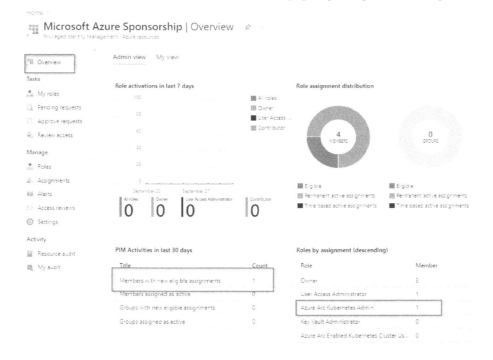

Figure 1.45 – Assignments on the Overview page

11. From **Assignments**, in the **Manage** section, you will see your assignment listed:

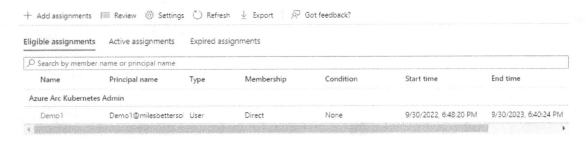

Figure 1.46 – Assignments

12. You should receive an email notification regarding this assignment; you can *update* or *remove* this assignment and create an *access review* for ongoing governance:

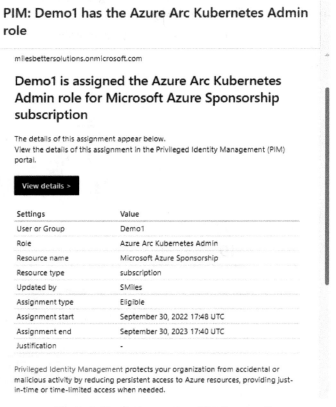

Figure 1.47 – Assignment notification email

With that, you have configured Privileged Identity Management. This concludes the hands-on tasks for this recipe.

How it works...

In this recipe, we looked at how to configure Privileged Identity Management. We assigned a user the Azure Arc Kubernetes Admin role.

See also

Should you require further information, you can refer to the following Microsoft Learn articles:

- Plan and implement privileged access: `https://learn.microsoft.com/en-us/training/modules/plan-implement-privileged-access`
- Azure Active Directory fundamentals documentation: `https://learn.microsoft.com/en-us/azure/active-directory/fundamentals`

2

Securing Azure Networks

In the previous chapter, we covered recipes that provided the foundation for securing Azure AD identities.

We should consider **Zero Trust** and **defense in depth** to be cornerstones of a cloud security strategy. We must consider the **network** as **untrusted** and **assume a breach**.

In this chapter, we build on those foundations and go through recipes that will equip us with the skills for securing Azure networks.

We will take a look at the protection of the network from the **Open Systems Interconnection (OSI)** model perspective and focus on solutions to protect **Layer 3** (**Network**), **Layer 4** (**Transport**), and **Layer 7** (**Application**).

By the end of this chapter, you will have gained valuable skills for securing Azure networks through the following recipes:

- Implementing network security groups
- Implementing Azure Firewall
- Implementing Azure Web Application Firewall
- Implementing Azure DDoS

Technical requirements

For this chapter, it is assumed that you have an Azure AD tenancy and an Azure subscription after completing the recipes in the previous chapter of this cookbook. If you skipped straight to this chapter, the information to create a new Azure AD tenancy and an Azure subscription for these recipes is included in the following list of requirements.

For this chapter, the following are required for the recipes:

- A device with a browser, such as Edge or Chrome, to access the Azure portal: `https://portal.azure.com`

- An **Azure AD tenancy** and **Azure subscription**; you may use an existing one or sign up for free: `https://azure.microsoft.com/en-us/free`
- An **Owner role** for the **Azure subscription**

Implementing network security groups

As part of an in-depth defense strategy, you should implement measures to protect your workload resources and filter network traffic between resources in your Azure virtual networks. **Network Security Groups (NSG)** can offer protection against lateral movement threats.

This recipe will teach you how to implement NSGs to protect your Azure virtual network virtual machine resources.

We will take you through creating a virtual network and a workload server virtual machine resource to protect. Then, we will walk through creating an NSG and apply it to the virtual network subnet where the test workload server virtual machine is located to demonstrate providing both allow and deny controls.

Getting ready

This recipe requires the following:

- A device with a browser, such as Edge or Chrome, to access the Azure portal: `https://portal.azure.com`
- You should sign in with an account that has the **Owner** or **Contributor** role for the **Azure subscription**
- An **Azure virtual machine** to use with this recipe; we will walk through creating this virtual machine as a *getting ready* task:
  - This will be created without an NSG attached to its network interface or the virtual machine's subnet
  - This will not have a public IP address associated with its network interface

Continue with the following *getting ready* task of creating a virtual machine for this recipe.

Getting ready task – creating a virtual machine

Perform the following steps:

1. In the search box in the *Azure portal*, type `virtual machines` and select **Virtual machines** from the listed **Services** results:

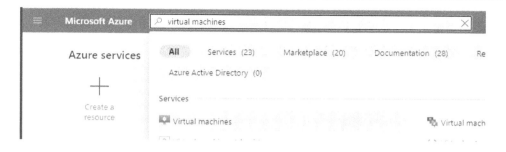

Figure 2.1 – Search for virtual machines

2. On the **Virtual machines** screen, click **+ Create** and then select **Azure virtual machine**:

Figure 2.2 – Virtual machines screen

3. From the **Basics** tab, under the **Project details** section, set the subscription as required.

4. For **Resource group**, select **Create new**.

5. Enter a name and click **OK**.

Figure 2.3 – Create a new resource group

6. Set the following:

    -   **Virtual machine name**: *Type a name*
    -   **Region**: *Select a region*
    -   **Availability options**: Select **No infrastructure redundancy required**
    -   **Security type**: Select **Standard**

- **Image**: Select **Windows Server 2019 Datacenter - Gen2**
- **Size**: *Leave the default (or set it as required to reduce recipe costs)*
- **Username** and **Password**: *Set as required*

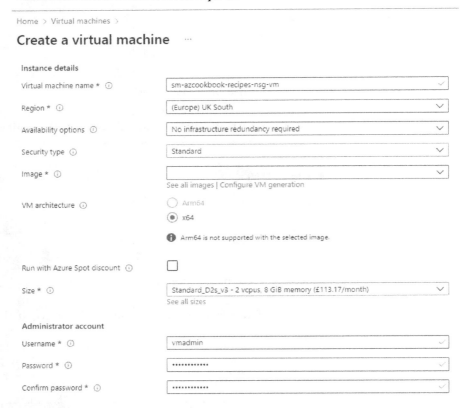

Figure 2.4 – Create a virtual machine

7. Set **Public inbound ports** to **None**.

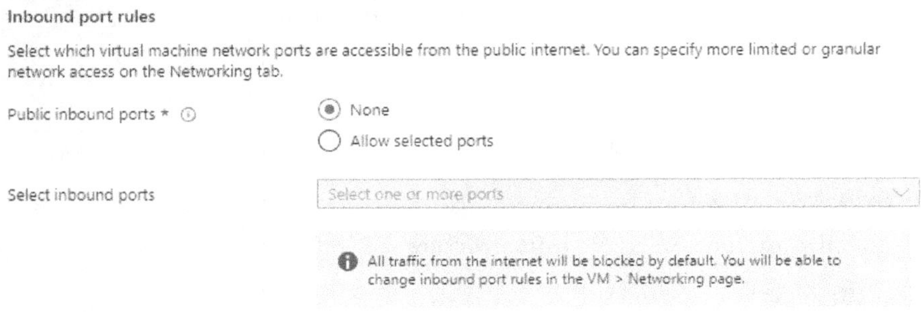

Figure 2.5 – Set Public inbound ports

8. Click **Next : Disks**, leave the *defaults*, then click **Next : Networking**.

9. From **virtual network**, click **Create new**.

10. On the **Create virtual network** screen, enter a name:

Figure 2.6 – Create virtual network

11. For **Address space**, leave the default.

12. Under **Subnets**, change the listed default subnet name from **default** to `Workload-Subnet`; then, click **OK**:

Figure 2.7 – Subnet settings

13. Set **Public IP** to **None** and **NIC network security group** to **None**, and tick to *enable* **Delete NIC when VM is deleted**:

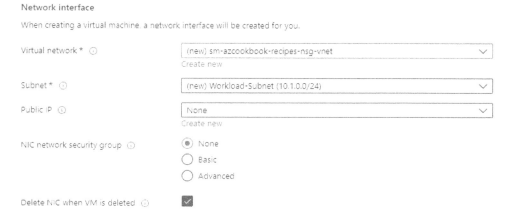

Figure 2.8 – Network interface settings

14. Click **Review + create**.

15. Click **Create** on the **Review + create** tab.

A notification will display that the resource deployment succeeded.

Figure 2.9 – Deployment completed notification

This getting ready task to create a virtual machine for this recipe is complete.

You are now ready to continue on to the main tasks of this recipe for implementing an NSG.

How to do it...

This recipe consists of the following tasks:

1. Create an NSG.

2. Associate the NSG with a subnet.

3. Add and test an inbound rule.

4. Add and test an outbound rule.

5. Clean up resources.

Task – creating an NSG

Perform the following steps:

1. In the search box in the Azure portal, type `network security groups` and select **Network security groups** from the listed **Services** results:

Figure 2.10 – Search for network security groups

2. On the **Network security groups** screen, click **+ Create**.

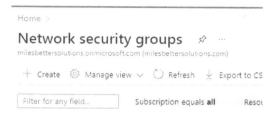

Figure 2.11 – Network security groups screen

3. From the **Basics** tab, set **Subscription** and **Resource group** to the same values that you set for your virtual machine in the *Getting ready* section.

Figure 2.12 – Project details

4. Set **Name** as required and **Region** to the same as you set for your virtual machine.
5. Click **Review + create**.
6. Click **Create** on the **Review + create** page.
7. Once the deployment is complete, click on **Go to resource**.

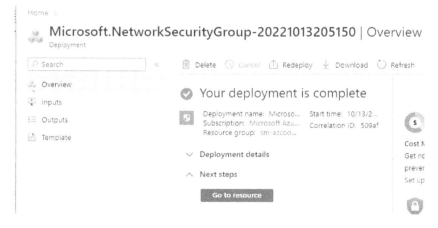

Figure 2.13 – Deployment completed notification

The task to create an NSG is complete. We will associate the created NSG with a subnet in the next task.

Task – associating the NSG with a subnet

Perform the following steps:

1. From the created NSG screen, click **Subnets** under the **Settings** section on the left menu.

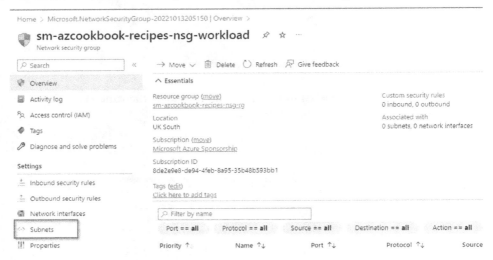

Figure 2.14 – NSG

2. Click **+ Associate** from the top toolbar.

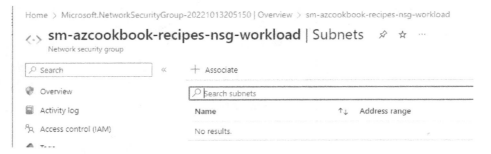

Figure 2.15 – Subnets screen

3. In the **Associate subnet** blade, select the virtual network and subnet of the virtual machine created in the previous task, then click **OK**.

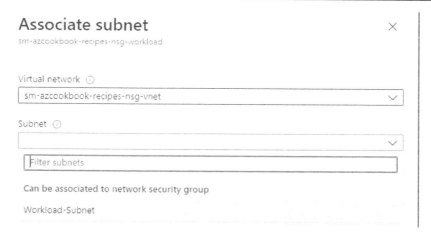

Figure 2.16 – Associate subnet

A notification will display that the changes were saved successfully.

The task to associate an NSG with a subnet is complete. In the next task, we add an inbound rule and test it.

Task – adding and testing an inbound rule

Perform the following steps:

1. Navigate to the NSG created in this recipe. Notice that all *inbound connections* are denied unless their source is **VirtualNetwork** or **Azure LoadBalancer**. We will address this in the next task.

2. In the **Network security group** blade, click **Inbound security rules** under **Settings**.

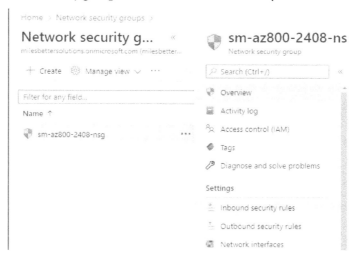

Figure 2.17 – Network security group

3. In the **Inbound security rules** blade, click **+ Add**.

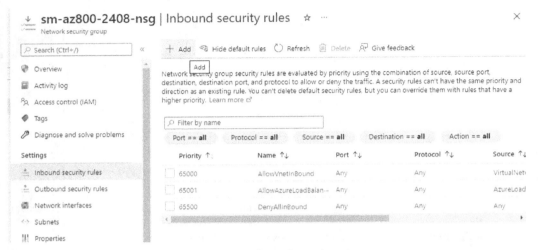

Figure 2.18 – Inbound security rules

4. Open a browser; in your chosen search engine, type what's my IP and note your IP address.

5. In the **Add inbound security rule** blade, set the following:

   - **Source**: Select **IP Addresses**

   - **Source IP addresses/CIDR ranges**: Set this to your IP address as discovered in step 4 of this task

   - **Source port ranges**: Leave the default of * (*asterisk symbol*)

   - **Destination**: Leave the default of **Any**

   - **Service**: Select **RDP**

   - **Action**: Ensure **Allow** is set

   - **Priority**: Leave the default of **100**

   - **Name**: Provide a name, such as AllowInbound_RDP_KnownIP

   - **Description**: Type as required

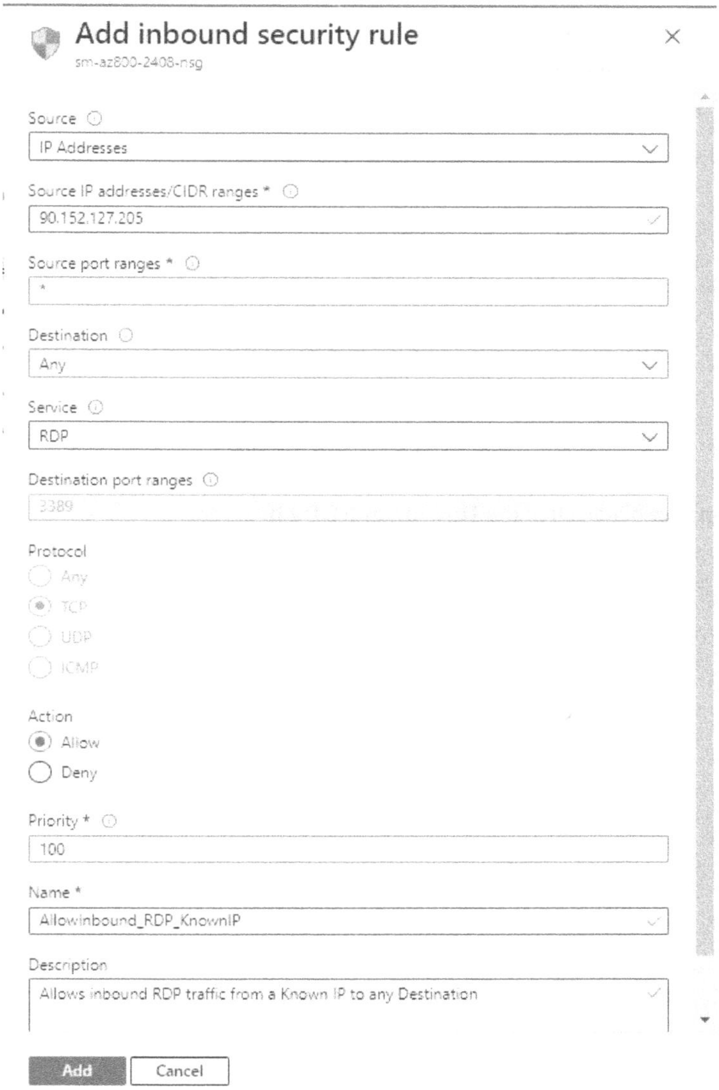

Add inbound security rule

sm-az800-2408-nsg

×

Source ⓘ

IP Addresses

Source IP addresses/CIDR ranges * ⓘ

90.152.127.205

Source port ranges * ⓘ

*

Destination ⓘ

Any

Service ⓘ

RDP

Destination port ranges ⓘ

3389

Protocol

◯ Any

◉ TCP

◯ UDP

◯ ICMP

Action

◉ Allow

◯ Deny

Priority * ⓘ

100

Name *

AllowInbound_RDP_KnownIP

Description

Allows inbound RDP traffic from a Known IP to any Destination

Add   Cancel

Figure 2.19 – Add inbound security rule

6.   Click **Add**. You will receive a notification that the rule was successfully created.

✅ Created security rule   ×

Successfully created security rule
'AllowInbound_RDP_KnownIP'.

Figure 2.20 – Security rule created

7. Navigate to the **Virtual machines** screen, click on your virtual machine, and from the **Overview** blade of your virtual machine, click on **Connect** and then **RDP**.

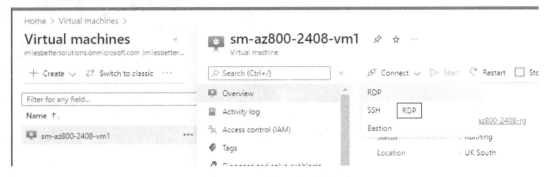

Figure 2.21 – Virtual machines

8. In the **Connect** blade, click the **Download RDP File** button.

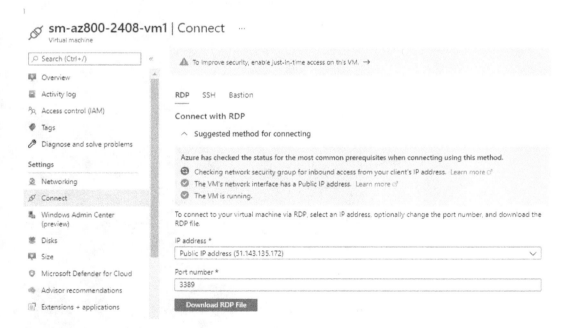

Figure 2.22 – Connect blade

9. Open the RDP file from where it was saved.

10. Click **Connect** to start an RDP session allowed by the *inbound rule* set in this task.

Figure 2.23 – Remote Desktop Connection

In this task, we added an inbound rule and completed our test of it. In the next task, we will create an outbound rule and test it.

Task – adding and testing an outbound rule

Perform the following steps:

1. In the **Virtual machine** blade, click **Networking** under **Settings**; you will see from the **Outbound port rules** tab that all outbound connections are allowed to the internet.

2. From the **Outbound port rules** tab, click **Add outbound port rule**.

3. In the **Add outbound security rule** blade, leave all options as the defaults apart from the following:

   - **Destination**: Select **Service Tag**

   - **Destination service tag**: Select **Internet**

   - **Destination port ranges**: Type the * symbol (*the asterisk symbol*)

   - **Action**: Ensure **Deny** is set

   - **Priority**: Leave the default value of **100**

- **Name**: Type a name, such as `DenyOutbound_Internet`
- **Description**: Type as required

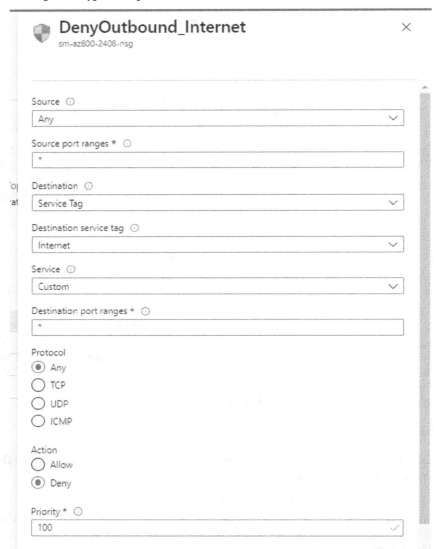

Figure 2.24 – Add security rule

4. Click **Add**. You will receive a notification that the rule was created successfully.

5. From the virtual machine, open a browser again and confirm you can no longer reach the internet by visiting a site such as `https://learn.microsoft.com`.

6. This time, you will see a message from the browser, such as **can't reach this page**.

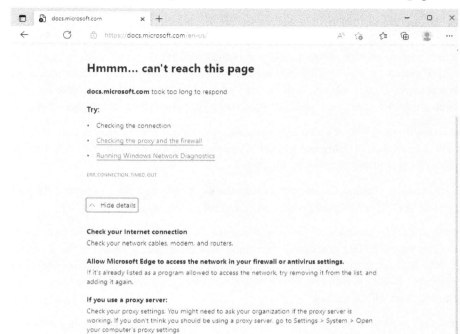

Figure 2.25 – Internet access denied

The task to create an outbound rule is complete. In the next task, we will clean up the resources created in this recipe.

Task – cleaning up resources

Perform the following steps:

1. In the search box in the Azure portal, type `resource groups` and select **Resource Groups** from the listed **Services** results.

2. From the **Resource groups** page, select the resource group we created for this recipe and click **Delete resource group**; this will delete all the resources created as part of this recipe.

Figure 2.26 – Delete resource group

The task to clean up the resources created in this recipe is complete.

How it works...

In this recipe, we looked at creating an NSG and associated it at the subnet level. We then added rules to allow **RDP access** on *port 3389* from a specified source IP address to a Windows Server virtual machine resource. We also added a rule to deny all outbound internet access. We also provided a step-by-step guide on creating a virtual network and a virtual machine to test these rules.

NSGs act as *traffic filters* and can be used to control traffic flow out of resources in a virtual network. *NSGs* contain rules like a traditional firewall that *allow* or *deny* inbound and outbound traffic; all traffic is *denied* unless *explicitly allowed* by a rule.

An **NSG** can only be associated with resources in the same subscription and region. **Azure Firewall** is better positioned for centralized protection across subscriptions and regions.

Each rule in an NSG is numbered; the lowest-numbered rule will be processed first. There is a set of default rules for each NSG that cannot be removed or modified. To override these rules, you add custom rules with a lower number, which are processed first. There is a deny-all final default rule that will be processed. That is, if your connection request is not explicitly allowed by a lower-number rule, then the connection will be denied.

Whether access is allowed or denied is based on the evaluation of the **five-tuple method**, which is based on the following *five data points*:

- Traffic **source**
- Traffic **source port**
- Traffic **destination**
- Traffic **destination port**
- Traffic **protocol**

The following diagram represents a simple example of traffic control with an NSG:

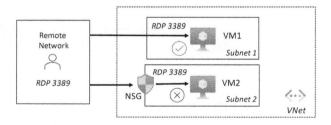

Figure 2.27 – NSG traffic control

In the preceding diagram, **no NSG** is *associated* with the *subnet* that **VM1** is connected to. This allows a connection with **RDP** on **port 3389** for a Windows virtual machine (*port 22 if Linux*). An attacker could brute-force attack this virtual machine using this unsecured management port.

VM2, however, has an *NSG* associated with the *VMs subnet* and, by default, will **deny** all inbound traffic unless **explicitly allowed**; the virtual machine management port will be protected against *brute-force attacks*.

To allow access, we must *explicitly* create a rule that specifies what connection traffic will be allowed.

If required, an **NSG** can be applied at the **subnet** or **network interface** level.

See also

Should you require further information, you can refer to the following Microsoft Learn articles:

- *Azure security baseline for Virtual Network*: `https://learn.microsoft.com/en-us/security/benchmark/azure/baselines/virtual-network-security-baseline`

- *Network security groups*: `https://learn.microsoft.com/en-us/azure/virtual-network/network-security-groups-overview`

- *How network security groups filter network traffic*: `https://learn.microsoft.com/en-us/azure/virtual-network/network-security-group-how-it-works`

- *Diagnose a virtual machine network traffic filter problem*: `https://learn.microsoft.com/en-us/azure/virtual-network/diagnose-network-traffic-filter-problem`

- *Tutorial: Log network traffic to and from a virtual machine using the Azure portal*: `https://learn.microsoft.com/en-us/azure/network-watcher/network-watcher-nsg-flow-logging-portal`

Implementing Azure Firewall

As part of our *defense-in-depth* strategy, we should implement measures to protect the perimeters of our Azure virtual networks. In environments with many distributed workload resources that need to communicate securely, we must ensure we protect these across many regions and subscriptions.

We must protect traffic entering our network from the internet (*North/South*), internal traffic from *spoke-to-spoke* virtual networks (*East/West*), and *cross-premises* hybrid or *partner edge* connections.

This recipe will teach you how to implement **Azure Firewall Premium** to protect your resources in an Azure virtual network.

We will take you through creating an Azure Firewall and policy, creating a default route, creating a workload server virtual machine for testing, and configuring and testing firewall rules.

Getting ready

This recipe requires the following:

- A device with a browser, such as Edge or Chrome, to access the Azure portal: `https://portal.azure.com`

- You should sign in with an account that has the **Owner** or **Contributor** role for the **Azure subscription**

- An **Azure virtual machine** to use with this recipe; we will walk through creating this virtual machine as a *getting ready* task:

  - This will be created without an NSG attached to its network interface or the virtual machine's subnet

  - This will not have a public IP address associated with its network interface

Continue with the following *getting ready* task of creating a virtual machine for this recipe.

Getting ready task – creating a workload server virtual machine

Perform the following steps:

1. In the search box in the Azure portal, type `virtual machines` and select **Virtual machines** from the listed **Services** results.

Figure 2.28 – Search for virtual machines

2. On the **Virtual machines** screen, click **+ Create** and then select **Azure virtual machine**:

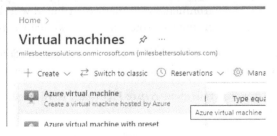

Figure 2.29 – Virtual machines

3. From the **Basics** tab, under the **Project details** section, select a subscription that will be used for resources in this recipe.

4. For **Resource group**, select **Create new**.

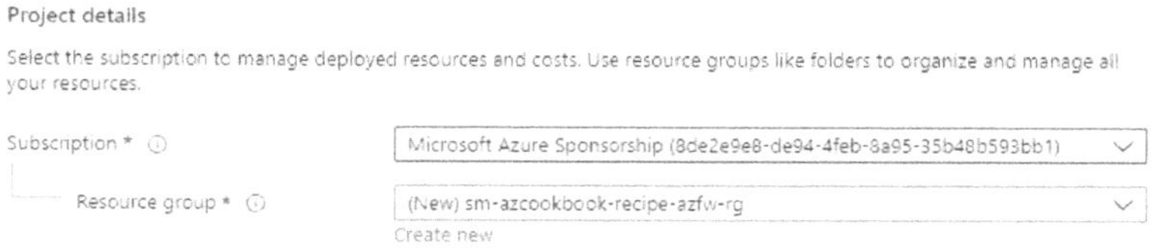

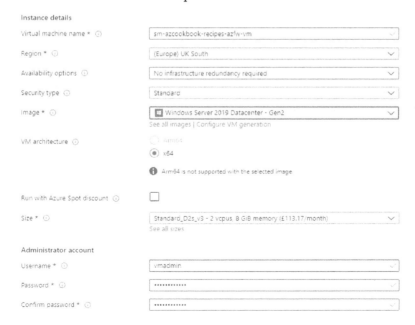

Figure 2.30 – Set project details

5. Set **Instance details** as follows:

   - **Virtual machine name**: Type a name

   - **Region**: Select the same region used to create the Azure Firewall

   - **Availability options**: Select **No infrastructure redundancy required**

   - **Security type**: Select **Standard**

   - **Image**: Select **Windows Server 2019 Datacenter - Gen2**

   - **Size**: Leave the default (or select as required to optimize costs)

   - Set **Username** and **Password** as required:

Figure 2.31 – Virtual machine settings

6. Set **Public inbound ports** to **None**:

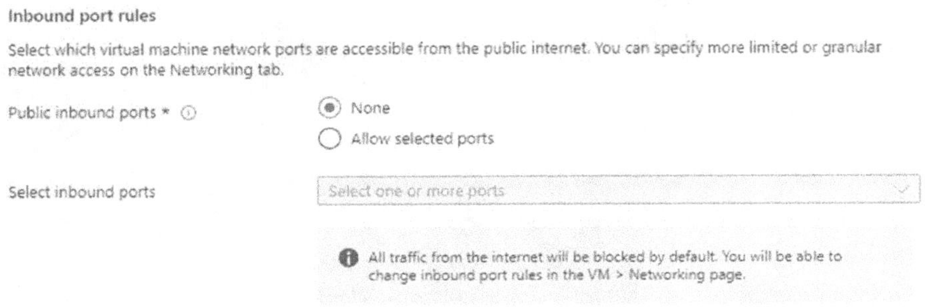

Inbound port rules

Select which virtual machine network ports are accessible from the public internet. You can specify more limited or granular network access on the Networking tab.

Public inbound ports * ⓘ   ◉ None
                                  ○ Allow selected ports

Select inbound ports   Select one or more ports   ⌄

            ⓘ All traffic from the internet will be blocked by default. You will be able to
               change inbound port rules in the VM > Networking page.

Figure 2.32 – Set Public inbound ports

7. Click **Next : Disks**, leave the defaults, then click **Next : Networking**.

8. In **Virtual network**, click **Create new**.

9. On the **Create virtual network** screen, enter a name:

Figure 2.33 – Create virtual machine

10. In the **Address space** section, under the **Address range** section, on the right-hand side of the provided default **Address range** entry, click the trash can icon to delete the default address range.

11. Type a new address range of 10.10.0.0/16.

12. In the **Subnets** section, for the subnet name, type WorkloadSubnet, and for **Address range**, type 10.10.1.0/24; then, click **OK**:

Figure 2.34 – Virtual networks

13. Set **Public IP** to **None** and **NIC network security group** to **None**, and tick to enable **Delete NIC when VM is deleted**:

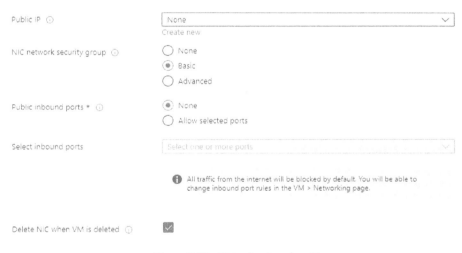

Figure 2.35 – Virtual network settings

14. Click **Review + create**.

15. Click **Create** on the **Review + create** tab.

16. On the screen that notifies you with **Your deployment is complete**, click on **Go to resource** to get ready for the next step in this task.

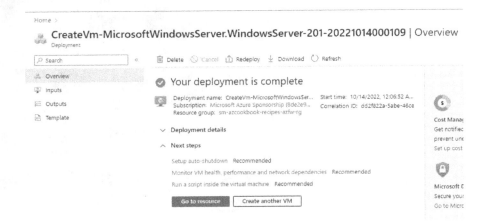

Figure 2.36 – Deployment complete

17. For the upcoming task of creating a **DNAT rule**, we will need to take note of the virtual machine's **private IP**; this can be found in the virtual machine's **Overview** blade under the **Networking** section of the **Properties** tab:

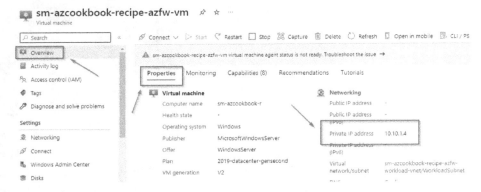

Figure 2.37 – Virtual machine blade

This *getting ready* task to create a virtual machine for this recipe is complete.

You are now ready to continue to the main tasks for this recipe for implementing Azure Firewall.

How to do it...

This recipe consists of the following tasks:

- Create an Azure Firewall.
- Create a workload server virtual machine for testing.

- Create virtual network peering.
- Create a **User-Defined Route (UDR)** to the workload subnet.
- Create a DNAT rule to allow RDP access to the workload server virtual machine.
- Create an application rule.
- Clean up resources.

Task – creating an Azure Firewall

Perform the following steps:

1. In the search box in the Azure portal, type `firewall` and select **Firewalls** from the listed **Services** results:

Figure 2.38 – Search for Azure Firewall

2. On the **Firewalls** screen, click **+ Create** or **Create firewall**:

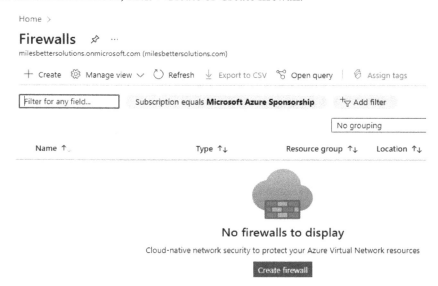

Figure 2.39 – Firewalls screen

3. From the **Basics** tab, under the **Project details** section, set **Subscription** as required:

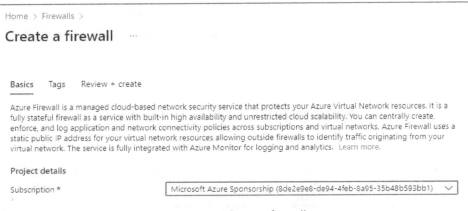

Figure 2.40 – Create a firewall

4. For **Resource group**, click **Create new**.

5. Enter a name and click **OK**:

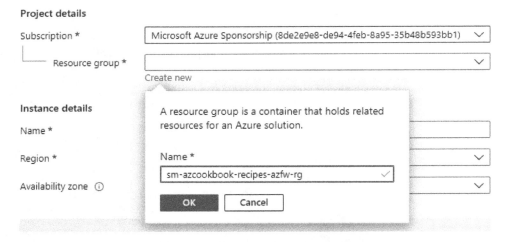

Figure 2.41 – Create a new resource group

6. Under the **Instance details** section, set a name and region as required and set **Availability zone** to **None**:

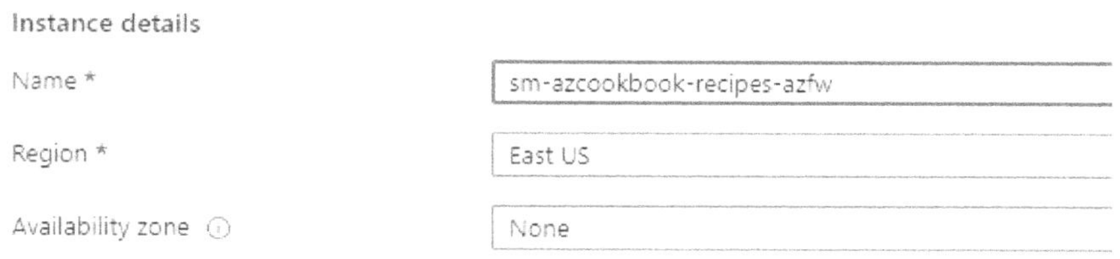

Instance details

Name *

    sm-azcookbook-recipes-azfw

Region *

    East US

Availability zone ⓘ

    None

Figure 2.42 – Create firewall settings

7. Ensure the **Firewall SKU** is set to **Premium**.

8. For **Firewall policy**, click **Add new**.

9. Type a policy name and region as required; ensure **Policy tier** is set to **Premium**. Then, click **OK**:

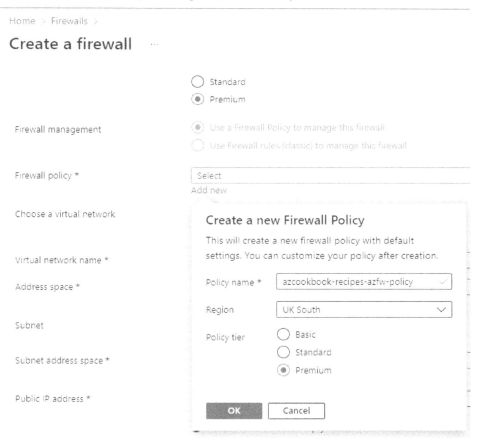

Figure 2.43 – Create a new Firewall Policy

10. For **Choose a virtual network**, leave it as **Create new**.

11. Type a virtual network name as required:

Figure 2.44 – Create a new virtual network

> **Note**
> We must ensure we do not overlap any address space used for the *workload server virtual machine* virtual network.

12. For **Address space**, type 10.0.0.0/16.

13. For **Subnet address space**, type 10.0.0.0/24:

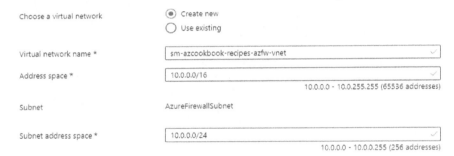

Figure 2.45 – Virtual network settings

14. For **Public IP address**, click **Add new**; type a name as required, then click **OK**:

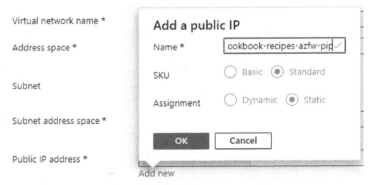

Figure 2.46 – Add a public IP

15. Click **Review + create**.

16. Click **Create** on the **Review + create** tab.

17. On the screen that notifies you with **Your deployment is complete**, click on **Go to resource** ready for the next step in this task:

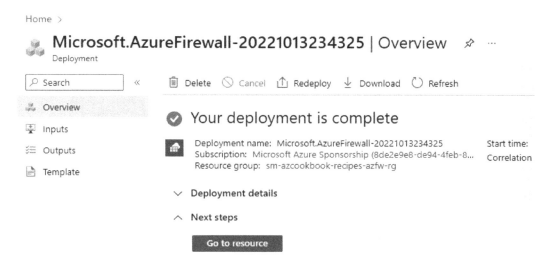

Figure 2.47 – Deployment complete

18. For the upcoming tasks of creating a **default route** and a **DNAT rule**, we will need to take note of the firewall's **public** and **private IPs**:

- The private IP can be found from the firewall's **Overview** blade:

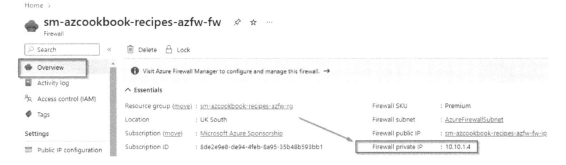

Figure 2.48 – Private IP

- The public IP can be found by clicking on the hyperlinked name for the public IP on the **Overview** blade, or by clicking on **Public IP configuration** under the **Settings** section:

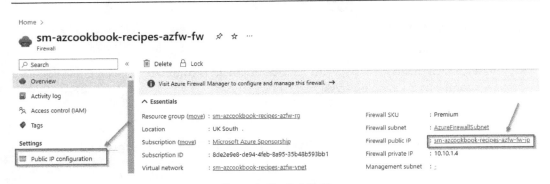

Figure 2.49 – Public IP

The task to create an Azure Firewall is complete. In the next task, we create peering between our two virtual networks.

Task – creating virtual network peering

Perform the following steps:

1. Navigate to the **Virtual networks** screen and you should see both virtual networks we created in this exercise; note that they are in the same subscription and region for this recipe.

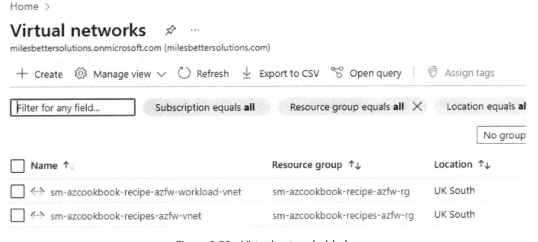

Figure 2.50 – Virtual networks blade

2. Click your Azure Firewall virtual network, **sm-azcookbook-recipes-azfw-vnet**.

3. From the **Virtual networks** blade, click **Peerings** under the **Settings** section of the left-hand menu:

Home >

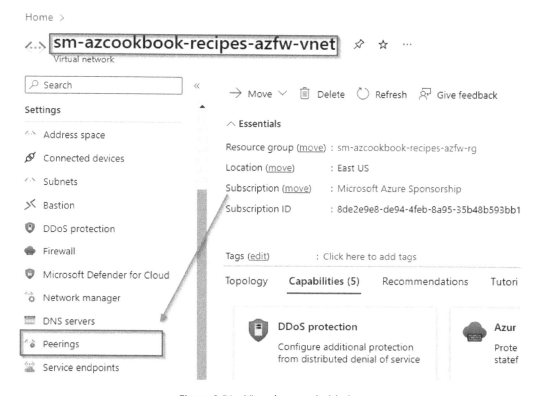

Figure 2.51 – Virtual networks blade

4. Click + **Add** from the top menu:

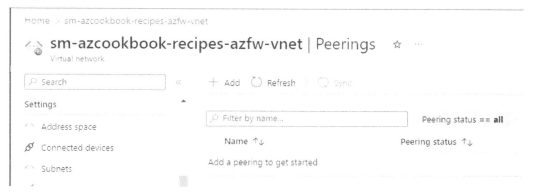

Figure 2.52 – Peerings blade

5. Enter a peering link name under the **This virtual network** section:

Home > Virtual networks > sm-azcookbook-recipes-azfw-vnet | Peerings >

Add peering   ...

sm-azcookbook-recipes-azfw-vnet

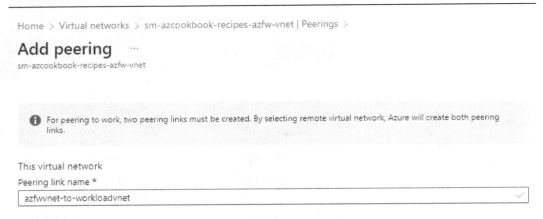

Figure 2.53 – Peering name for this virtual network

6. Enter a peering link name under the **Remote virtual network** section:

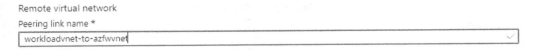

Figure 2.54 – Peering name for remote virtual network

7. Select the same subscription used for previous tasks in this recipe; for **Virtual network**, select
    the workload virtual network we created that contains our workload server virtual machine,
    then click **Add**:

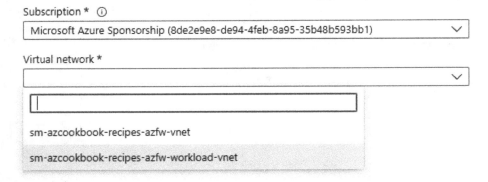

Figure 2.55 – Add peering

You will see a notification that the peering was successfully added:

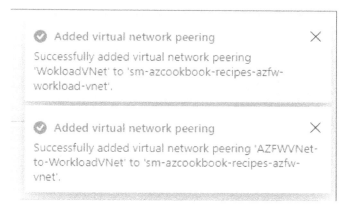

Figure 2.56 – Peering added

The task to create peering between virtual networks is complete. In the next task, we create a UDR.

Task – creating a user-defined route

Perform the following steps:

1. In the search box in the Azure portal, type `route table` and select **Route tables** from the listed **Services** results:

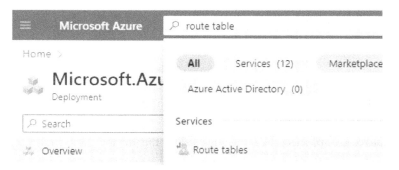

Figure 2.57 – Search for route table

2. On the **Route tables** screen, click **+ Create** or **Create route table**:

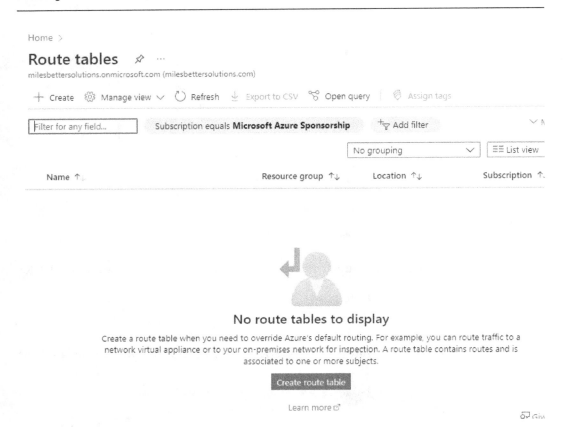

Figure 2.58 – Route tables screen

3. From the **Basics** tab, in the **Project details** section, select the same subscription used for the previous tasks in this recipe.

4. For **Resource group**, select the same one used for the previous tasks in this recipe:

Figure 2.59 – Set project details

5. Select the same region as used for the previous tasks in this recipe; then, type a name as required:

Figure 2.60 – Set instance details

6. Click **Review + create**.

7. Click **Create** on the **Review + create** tab.

8. Click **Go to resource** on the **Deployment** screen that notifies you that the deployment is complete:

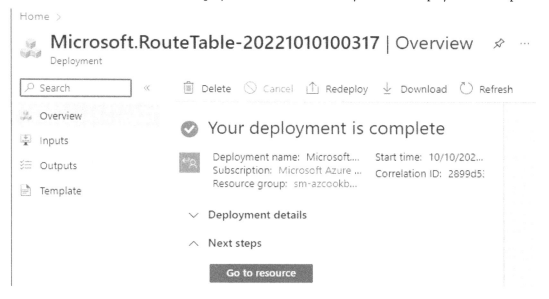

Figure 2.61 – Deployment completed

9. Click **Subnets** under the **Settings** section on the **Route table** screen, then click **+ Associate**:

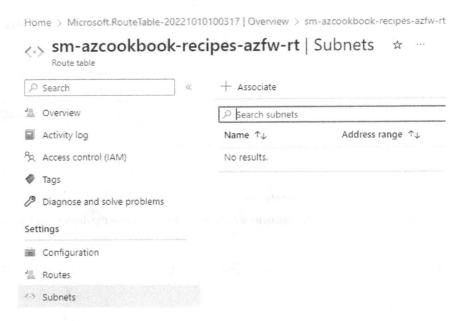

Figure 2.62 – Associate route table

10. For **Virtual network** on the **Associate subnet** blade, select your workload virtual network created for this recipe; then, select **Workload-Subnet** from the **Subnet** list and click **OK**:

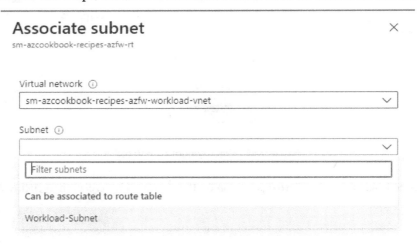

Figure 2.63 – Associate subnet

11. Select **Routes** under the **Settings** section on the **Route table** screen and click **+ Add**:

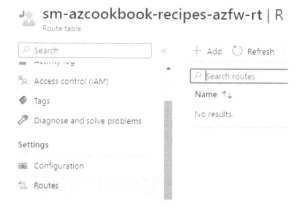

Figure 2.64 – Add route

12. In the **Add route** blade, add a route name to represent the default gateway for the workload subnet and select **IP Addresses** for **Address prefix destination**:

Figure 2.65 – Add route

13. For **Destination IP addresses/CIDR ranges**, type 0.0.0.0/0.

14. For **Next hop type**, select **Virtual appliance**:

Figure 2.66 – Route table configuration

15. For **Next hop address**, type Azure Firewall's private IP that we noted at the beginning of this task; for reference, this can be found in the **Overview** blade of your firewall. Then, click **Add**:

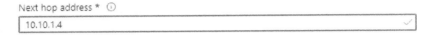

Figure 2.67 – Route table configuration

16. A notification will display that you successfully added a route:

Figure 2.68 – Added route

The task to create a UDR is complete. In the next task, we create a **Destination Network Address Translation (DNAT)** rule.

Task – creating a DNAT rule

Perform the following steps:

1. Navigate to your created Azure Firewall in the Azure portal, and in the **Overview** blade of your firewall, under **Firewall policy**, click your policy to open the **Firewall Policy** screen:

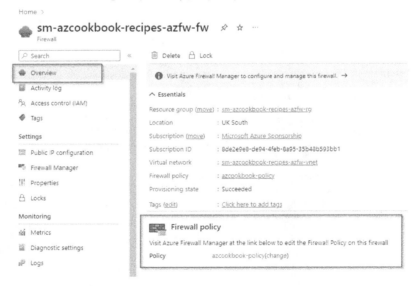

Figure 2.69 – Azure Firewall

2. On the **Firewall Policy** screen, click **DNAT rules** under the **Settings** section:

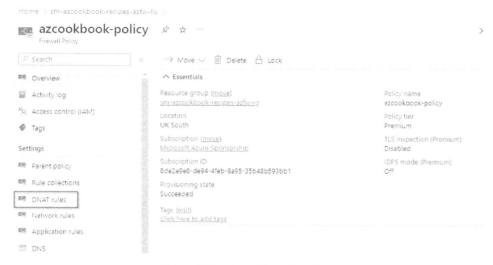

Figure 2.70 – Azure Firewall policy

3. Click + **Add a rule collection**:

Figure 2.71 – Add rule

4. On the **Add a rule collection** blade, set the following:

    •   Type a name as required for the rule collection

    •   For **Rule collection type**, ensure **DNAT** is selected

    •   For **Priority**, type 2 0 0

    •   For **Rule collection group**, select **DefaultDnatRuleCollectionGroup**:

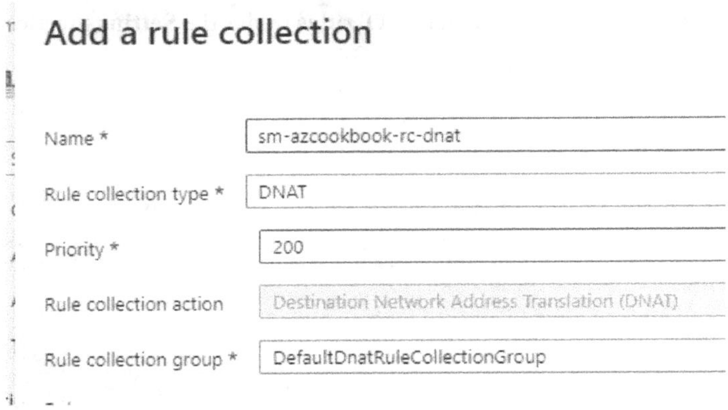

Add a rule collection

Name *	sm-azcookbook-rc-dnat
Rule collection type *	DNAT
Priority *	200
Rule collection action	Destination Network Address Translation (DNAT)
Rule collection group *	DefaultDnatRuleCollectionGroup

Figure 2.72 – Add a rule collection

5.   In the **Rules** section, set the following:

   - Type a name as required for the first rule in the rule collection
   - For **Source type**, ensure **IP Address** is selected
   - For **Source**, type *
   - For **Protocol**, select **TCP**
   - For **Destination Ports**, type 3389
   - For **Destination Type**, ensure **IP Address** is selected
   - For **Destination**, enter the **Firewall public IP address** we noted in an earlier task in this recipe
   - For **Translated address**, enter the **workload server virtual machine private IP address** we noted in an earlier task in this recipe
   - For **Translated port**, type 3389

You should now have a rule that looks like the following:

Name *	Source type	Source	Protocol *	Destination Ports *	Destination Type *
am-azcookbook-r...	IP Address	*	TCP	3389	IP Address

Destination Type *	Destination *	Translated address *	Translated port *	
IP Address	20.68.27.149	10.10.1.4	3389	🗑 •••

Figure 2.73 – Firewall rule

6. Click **Add**. You will be notified that the rule collection was successfully added in a few minutes:

Figure 2.74 – Added rule collection

The task to create a DNAT rule is complete. In the next task, we create an application rule.

Task – creating an application rule

Perform the following steps:

1. Navigate to the created Azure Firewall in the Azure portal, and in the **Overview** blade of your firewall, under the **Firewall policy** section, click your Firewall policy to open the **Firewall Policy** screen:

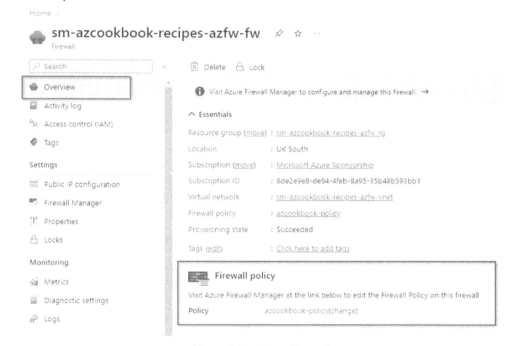

Figure 2.75 – Azure Firewall

2.   On the **Firewall Policy** screen, click **Application rules** under the **Settings** section:

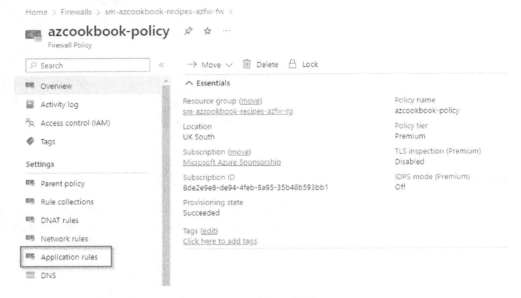

Figure 2.76 – Firewall Policy

3.   Click + **Add a rule collection**.

4.   On the **Add a rule collection** blade, set the following:

- Type a name as required for the rule collection

- For **Rule collection type**, ensure **Application** is selected

- For **Priority**, type 200

- For **Rule collection action**, select **Allow**

- For **Rule collection group**, select **DefaultApplicationRuleCollectionGroup**:

Add a rule collection

Name *	azcookbook-rc-app
Rule collection type *	Application
Priority *	200
Rule collection action	Allow
Rule collection group *	DefaultApplicationRuleCollectionGroup

Figure 2.77 – Add rule collection

5. In the **Rules** section, set the following:

- Type a name as required for the first rule in the rule collection

- For **Source type**, ensure **IP Address** is selected

- For **Source**, type *

- For **Protocol**, select **https**

- For **Destination Type**, select **FQDN**

For **Destination**, type learn.microsoft.com

- You should now have a rule that looks like the following example:

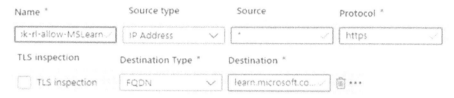

Figure 2.78 – Firewall rule

6. Click **Add**. You will be notified that the rule collection was successfully added in a few minutes:

Figure 2.79 – Added rule collection

The task to create an application rule is complete. In the next tasks, we test the firewall rules.

Task – testing the firewall DNAT rule

Perform the following steps:

1. Open **Remote Desktop Connection** and set it to the firewall's public IP address:

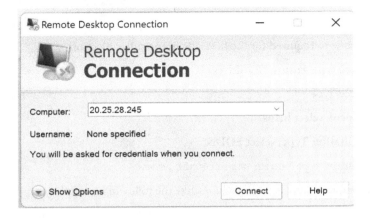

Figure 2.80 – Remote Desktop Connection

2. You will be prompted with the workload server virtual machine login credentials:

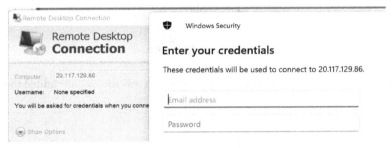

Figure 2.81 – Access to workload server virtual machine

3. Log on to the workload server virtual machine with the credentials used when it was created:

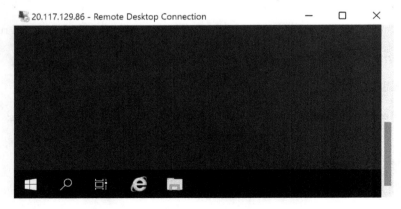

Figure 2.82 – Logged on to workload server virtual machine via RDP

The task to test the firewall DNAT rule is complete. In the next task, we test the firewall application rule.

Task – testing the firewall application rule

Perform the following steps:

1. From the desktop of the workload server virtual machine, open a browser and go to `www.google.com` or `www.microsoft.com`, and ensure your connection is blocked:

```
Action: Deny. Reason: No rule matched. Proceeding with default action.
```

Figure 2.83 – Access denied

2. From **Server Manager**, set **IE Enhanced Security Configuration** to **Off**.

3. From a browser, go to `https://learn.microsoft.com` and ensure your connection is allowed:

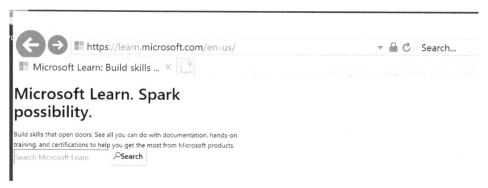

Figure 2.84 – Access allowed

The task to test the firewall application rule is complete. In the next task, we clean up the resources created in this recipe.

Task – cleaning up resources

Perform the following steps:

1. In the search box in the Azure portal, type `resource groups` and select **Resource Groups** from the listed **Services** results.

2. On the **Resource groups** page, select the resource group that we created for this recipe and click **Delete resource group**. This will delete all the resources created as part of this recipe:

Figure 2.85 – Delete resource group

The task to clean up the resources created in this recipe is complete.

How it works...

In this recipe, we looked at implementing an Azure Firewall instance and created a virtual machine to use as a workload server to test our rules.

We implemented a **hub-and-spoke** virtual network topology. Azure Firewall was deployed into the **Hub** virtual network. The virtual machine server was deployed into a **Spoke** virtual network (within the same region) and was connected via **virtual network peering**.

The Azure Firewall network topology we created is represented in the following diagram:

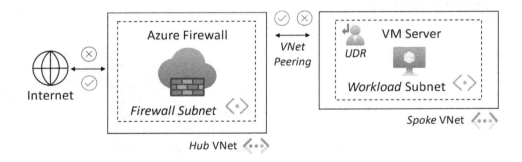

Figure 2.86 – Azure Firewall deployment topology

We created UDRs through a **route table**, associated it with the *workload subnet*, and created a *new route* to direct traffic with a destination of 0.0.0.0/0 to **Azure Firewall** as the *next hop*.

As part of the recipe for configuring Azure Firewall, we created a **DNAT rule** that allowed *RDP access* to our test virtual machine. We created outbound *application rules* to block and allow specified URLs.

Azure Firewall is a fully stateful Microsoft-managed firewall and an Azure-hosted *firewall-as-a-service* platform solution. It provides *Layers 3-7* centralized control policies and consumes threat-intelligence feeds directly from Microsoft's cybersecurity platforms, providing real-time insights and protection.

Segmentation of networks can be used with *hub-and-spoke* network topologies. The traffic flow and control from the spoke networks are provided through UDRs.

Azure Firewall can provide the following capabilities:

- **Intrusion Prevention System (IDS)**
- **Transport Layer Security (TLS)** inspection
- **Uniform Resource Locator (URL)** filtering
- **Source Network Address Translation (SNAT)**
- **Destination Network Address Translation (DNAT)**

It is important to note that **TLS inspection** is only supported for **outbound** (*North*) and **lateral** (*East/West*) traffic, that is, an inspection of traffic from an internal Azure-hosted client to the internet and sent from within Azure and to/from on-premises.

Azure Firewall Manager provides centralized policy configuration and management for multiple Azure Firewall instances.

See also

Should you require further information, you can refer to the following Microsoft Learn articles:

- *Azure Firewall documentation*: https://learn.microsoft.com/en-us/azure/firewall

- *Azure Firewall FAQ*: https://learn.microsoft.com/en-us/azure/firewall/firewall-faq

- *Hub-spoke network topology in Azure*: https://learn.microsoft.com/en-us/azure/architecture/reference-architectures/hybrid-networking/hub-spoke

- *Azure Firewall web categories*: https://learn.microsoft.com/en-us/azure/firewall/web-categories

- *FQDN tags overview*: https://learn.microsoft.com/en-us/azure/firewall/fqdn-tags

Implementing Azure Web Application Firewall

As we continue with our defense-in-depth strategy, we should look at the different types of traffic on the network, their protocols, and their direction, such as *inbound/outbound* and *lateral* traffic flows; this can be referred to as *north/south* and *east/west* traffic.

We should evaluate the most appropriate defense mechanism based on our desired outcomes. If we allow any *HTTP(s)* protocols into our Azure networks, such as to allow access to web applications, we need to implement measures to protect against *Layer 7 web protocol* attacks, such as *cross-site scripting* and *SQL injection*.

This outcome can be achieved by implementing a *Layer 7* **Web Application Firewall** (**WAF**), rather than a *Layer 4* **network firewall**.

It is important to note that a traditional *Layer 4 network firewall* will not offer protection against these inbound *Layer 7* attacks; an **Intrusion Detection and Prevention System** (**IDPS**) solution will also be ineffective in detecting attacks in encrypted traffic.

Regarding inspecting encrypted traffic, we saw in the previous section that *Azure Firewall* could provide *TLS inspection*; however, this only supports **outbound** (*North*) internet traffic and **lateral** (*East/West*) traffic that stays within the Azure network or traverses cross-premises. **Inbound** (*South*) **TLS** inspection needs a **WAF**.

In this section, we will look at a recipe to implement a WAF using the *Azure Application Gateway* service to protect exposed *HTTP(s)* web services from *Layer 7* web protocol attacks.

Getting ready

This recipe requires the following:

- A device with a browser, such as Edge or Chrome, to access the Azure portal: `https://portal.azure.com`

- You should sign in with an account that has the **Owner** or **Contributor** role for the **Azure subscription**

- An **Azure virtual machine**; we will walk through creating this virtual machine as a *getting ready task*:

  - This will be created without an NSG attached to its network interface or the virtual machine's subnet

  - This will not have a public IP address associated with its network interface

Please continue with the following *getting ready* task of creating a workload server virtual machine with IIS installed to act as our application server for testing access through the application gateway.

Getting ready task – creating a workload server virtual machine with IIS to test access

Perform the following steps:

1. In the search box in the Azure portal, type `virtual machines` and select **Virtual machines** from the listed **Services** results:

Figure 2.87 – Search for virtual machines

2. On the **Virtual machines** screen, click **+ Create** and then select **Azure virtual machine**:

Figure 2.88 – Virtual machines screen

3. From the **Basics** tab, under the **Project details** section, set the subscription as required.

4. Select the resource group we will use to create the application gateway in this recipe:

Figure 2.89 – Set project details

5. Set the following:

- **Virtual machine name**: Type a name
- **Region**: Select a region
- **Availability options**: Select **No infrastructure redundancy required**
- **Security type**: Select **Standard**
- **Image**: Select **Windows Server 2019 Datacenter - Gen2**
- **Size**: Leave the default (*or set it as required to reduce recipe costs*)
- Set **Username** and **Password** as required:

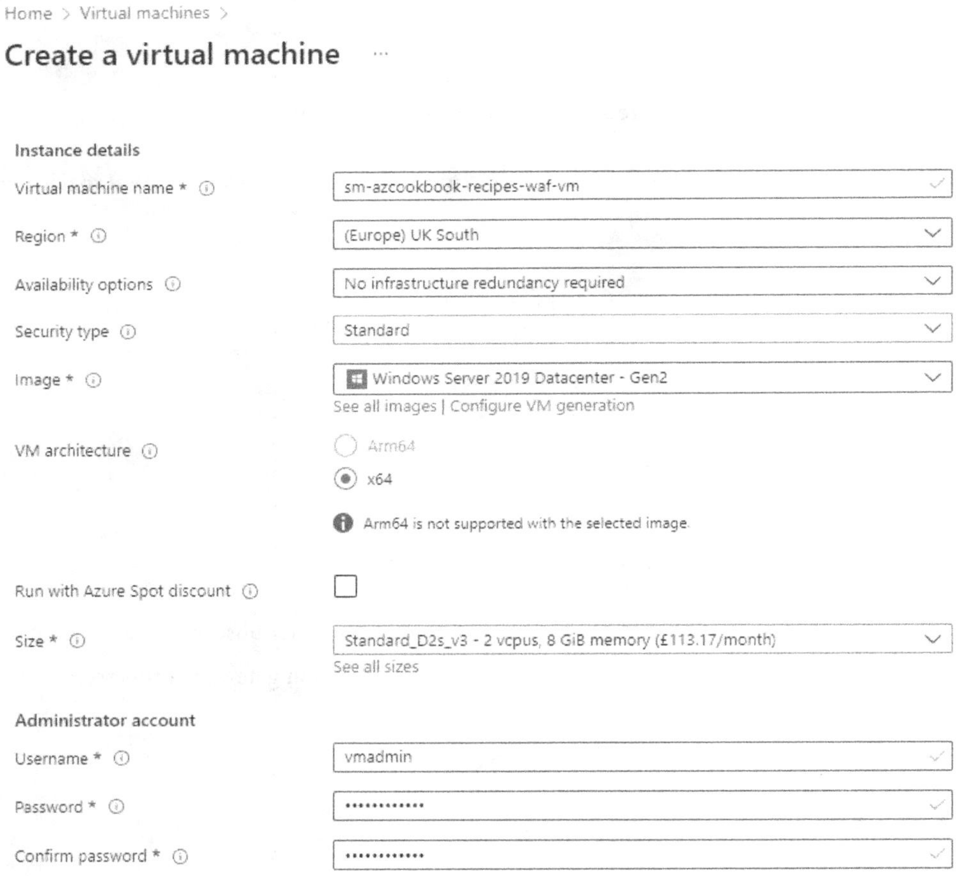

Home > Virtual machines >

Create a virtual machine   ...

Instance details

Virtual machine name * ⓘ	sm-azcookbook-recipes-waf-vm
Region * ⓘ	(Europe) UK South
Availability options ⓘ	No infrastructure redundancy required
Security type ⓘ	Standard
Image * ⓘ	▓ Windows Server 2019 Datacenter - Gen2
	See all images | Configure VM generation
VM architecture ⓘ	◯ Arm64
	◉ x64
	❶ Arm64 is not supported with the selected image.
Run with Azure Spot discount ⓘ	☐
Size * ⓘ	Standard_D2s_v3 - 2 vcpus, 8 GiB memory (£113.17/month)
	See all sizes

Administrator account

Username * ⓘ	vmadmin
Password * ⓘ	••••••••••••
Confirm password * ⓘ	••••••••••••

Figure 2.90 – Create a virtual machine

6. Set **Public inbound ports** to **None**:

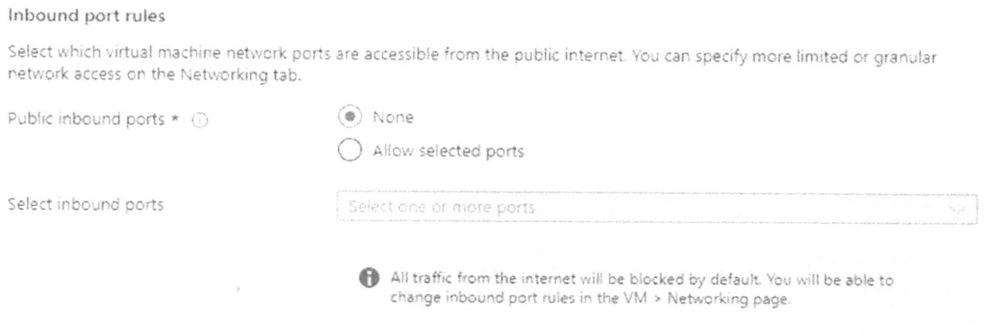

Inbound port rules

Select which virtual machine network ports are accessible from the public internet. You can specify more limited or granular network access on the Networking tab.

Public inbound ports * ⓘ   ◉ None

                               ◯ Allow selected ports

Select inbound ports   Select one or more ports   ⌄

                         ❶ All traffic from the internet will be blocked by default. You will be able to
                            change inbound port rules in the VM > Networking page.

Figure 2.91 – Set inbound port rules

7. Click **Next : Disks**, leave the defaults, then click **Next : Networking**.

8. Select the virtual network we created earlier in this recipe when we created the application gateway.

9. Select **WorkloadSubnet** for this virtual network:

Network interface

When creating a virtual machine, a network interface will be created for you.

Virtual network * ⓘ   smazcookbook-recipes-waf-vnet   ⌄
                             Create new

Subnet * ⓘ   WorkloadSubnet (10.0.1.0/24)   ⌄

                             Filter subnets

Public IP ⓘ   WorkloadSubnet (10.0.1.0/24)

NIC network security group ⓘ

Figure 2.92 – Configure the network interface

10. Set **Public IP** to **None**.

11. Set **NIC network security group** to **None**.

12. Tick to enable **Delete NIC when VM is deleted**.

13. Click **Review + create**.

14. Click **Create** on the **Review + create** tab.

A notification will display that the resource deployment has succeeded:

Figure 2.93 – Deployment completed

15. To install IIS on the virtual machine, open **Cloud Shell** from the top navigation bar of the Azure portal:

Figure 2.94 – Launch Cloud Shell

16. If this is the first time you have run Cloud Shell, you will be prompted to create a storage account; select your subscription, then click **Create storage**:

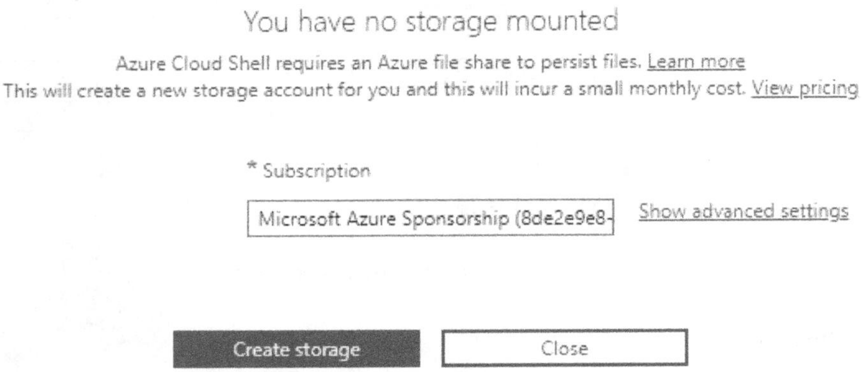

You have no storage mounted

Azure Cloud Shell requires an Azure file share to persist files. Learn more
This will create a new storage account for you and this will incur a small monthly cost. View pricing

* Subscription

Microsoft Azure Sponsorship (8de2e9e8-   Show advanced settings

Create storage   Close

Figure 2.95 – Create Cloud Shell storage

17. In **Cloud Shell**, select **PowerShell**:

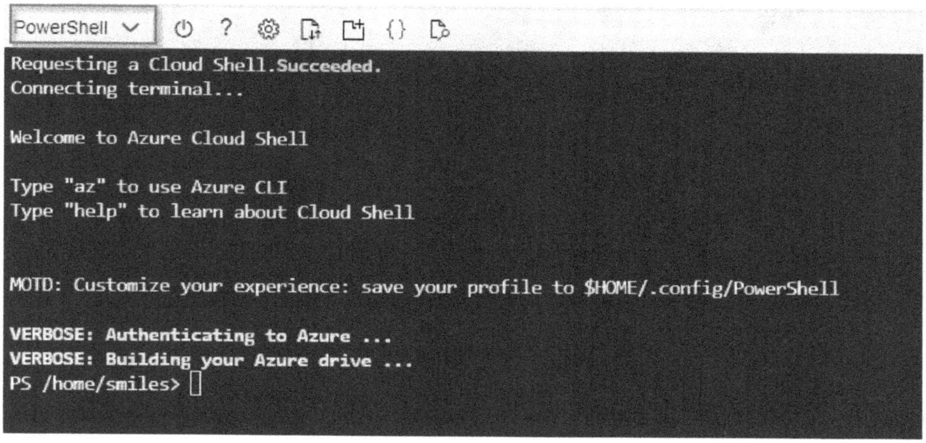

Figure 2.96 – Launch PowerShell

18. Using your environment values for `ResourceGroupName`, `VMName`, and `Location`, run the following PowerShell command:

```
Set-AzVMExtension `
    -ResourceGroupName sm-azcookbook-recipes-waf-rg `
    -ExtensionName IIS `
    -VMName sm-azcookbook-recipes-waf-vm `
    -Publisher Microsoft.Compute `
    -ExtensionType CustomScriptExtension `
    -TypeHandlerVersion 1.4 `
    -SettingString '{"commandToExecute":"powershell
Add-WindowsFeature Web-Server; powershell Add-Content
-Path \"C:\\inetpub\\wwwroot\\Default.htm\" -Value
$($env:computername)"}' `
    -Location WestEurope
```

The preceding code is represented in the following figure:

```
PS /home/smiles> Set-AzVMExtension `
>>    -ResourceGroupName sm-azcookbook-recipes-waf-rg `
>>    -ExtensionName IIS `
>>    -VMName sm-azcookbook-recipes-waf-vm `
>>    -Publisher Microsoft.Compute `
>>    -ExtensionType CustomScriptExtension `
>>    -TypeHandlerVersion 1.4 `
>>    -SettingString '{"commandToExecute":"powershell Add-WindowsFeature Web-Server; powershell Add-Content -Path
\"C:\\inetpub\\wwwroot\\Default.htm\" -Value $($env:computername)"}' `
>> -Location WestEurope


RequestId IsSuccessStatusCode StatusCode ReasonPhrase
--------- ------------------- ---------- ------------
          True                OK OK
```

Figure 2.97 – Install web server feature

This *getting ready* task was for creating a workload server virtual machine with IIS installed to act as our application server for testing access through the application gateway.

You are now ready to continue to the main tasks for this recipe for implementing a WAF-enabled application gateway.

How to do it...

This recipe consists of the following tasks:

- Create an application gateway with WAF enabled.
- Add a server to the backend pool.
- Test the application gateway with WAF enabled.
- Clean up resources.

Task – creating an application gateway with WAF enabled

Perform the following steps:

1. In the search box in the Azure portal, type application gateway and select **Application gateways** from the listed **Services** results:

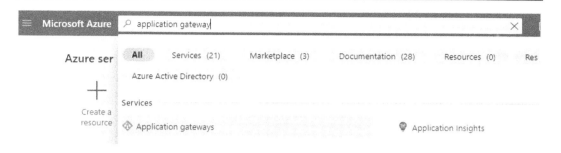

Figure 2.98 – Search for application gateway

2. On the **Application Gateway** screen, click + **Create** or **Create application gateway**:

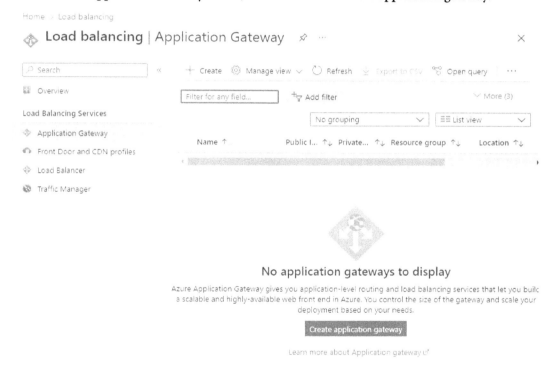

Figure 2.99 – Application Gateway screen

3. From the **Basics** tab, under the **Project details** section, set **Subscription** as required.

4. For **Resource group**, click **Create new**.

5. Enter a name and click **OK**:

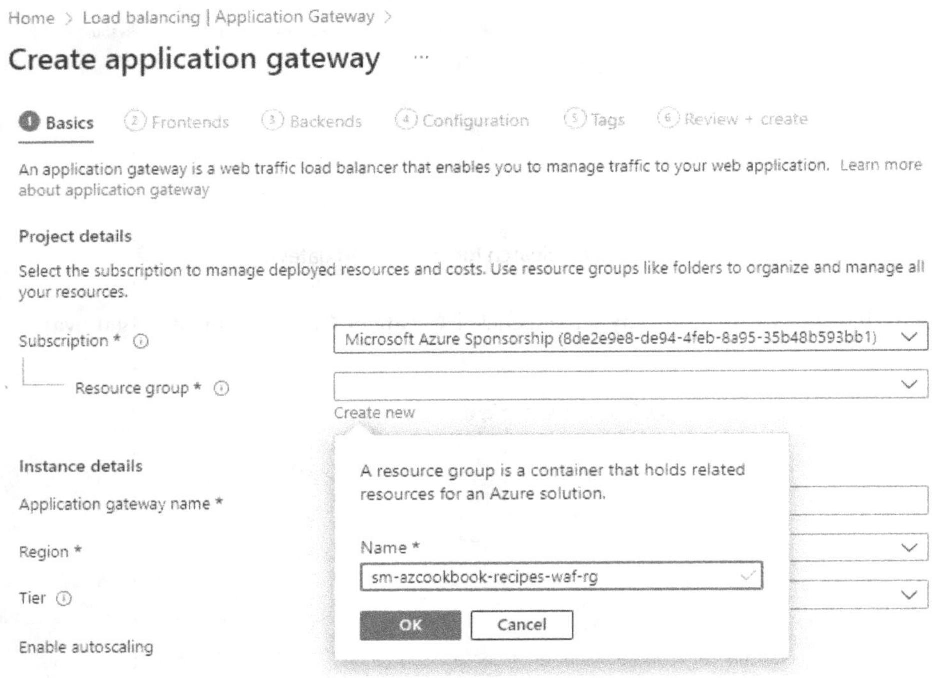

Home > Load balancing | Application Gateway >

Create application gateway   ...

① **Basics**   ② Frontends   ③ Backends   ④ Configuration   ⑤ Tags   ⑥ Review + create

An application gateway is a web traffic load balancer that enables you to manage traffic to your web application. Learn more about application gateway

Project details

Select the subscription to manage deployed resources and costs. Use resource groups like folders to organize and manage all your resources.

Subscription * ⓘ | Microsoft Azure Sponsorship (8de2e9e8-de94-4feb-8a95-35b48b593bb1) ∨ |

　　　Resource group * ⓘ | ∨ |
                            Create new

Instance details   A resource group is a container that holds related
                                        resources for an Azure solution.
Application gateway name *

Region *   Name *   ∨

Tier ⓘ | sm-azcookbook-recipes-waf-rg   ∨ | ∨

Enable autoscaling   [OK]   [Cancel]

Figure 2.100 – Set project details

6. Set the following information under the **Instance details** section:

- **Application gateway name**: Enter as required

- **Region**: Enter the same region selected for the workload virtual machine we created

- **Tier**: Select **WAF V2**

- **Enable autoscaling**: Leave the default

- **Minimum instance count**: Leave the default

- **Maximum instance count**: Leave the default

- **Availability zone**: Leave the default

- **HTTP2**: Leave the default

- **WAF Policy**: Click **Create new**; in the **Create Web Application Firewall Policy** blade, enter a name and select **Add Bot Protection** if required:

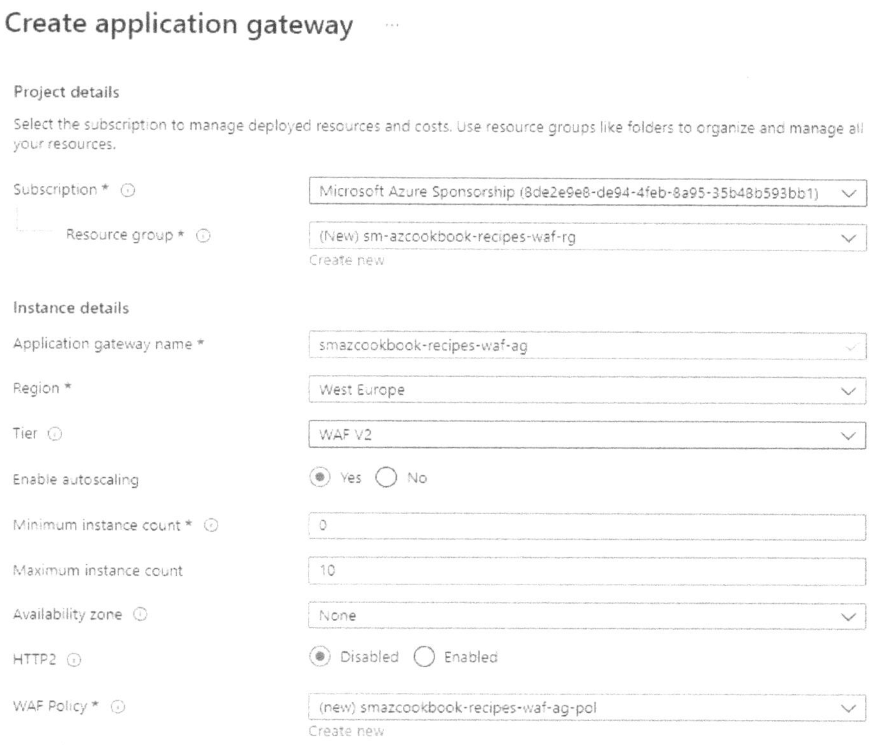

Create application gateway ...

Project details

Select the subscription to manage deployed resources and costs. Use resource groups like folders to organize and manage all your resources.

Subscription * ⓘ
> Microsoft Azure Sponsorship (8de2e9e8-de94-4feb-8a95-35b48b593bb1)   ∨

Resource group * ⓘ
> (New) sm-azcookbook-recipes-waf-rg   ∨
> Create new

Instance details

Application gateway name *
> smazcookbook-recipes-waf-ag   ∨

Region *
> West Europe   ∨

Tier ⓘ
> WAF V2   ∨

Enable autoscaling
> ⦿ Yes   ◯ No

Minimum instance count * ⓘ
> 0

Maximum instance count
> 10

Availability zone ⓘ
> None   ∨

HTTP2 ⓘ
> ⦿ Disabled   ◯ Enabled

WAF Policy * ⓘ
> (new) smazcookbook-recipes-waf-ag-pol   ∨
> Create new

Figure 2.101 – Create application gateway

7. In the **Configure virtual network** section, click **Create new** to create a new virtual network for this recipe.

8. On the **Create virtual network** screen, enter a name.

9. In the **ADDRESS SPACE** section, leave the default.

10. In the **SUBNETS** section, set the following:

    - *Rename* the default subnet to WAF-Subnet.

    - In the second row of the table, type a subnet name of WorkloadSubnet, and for **Address range**, type 10.0.1.0/24. Click **OK**:

SUBNETS

The subnet's address range in CIDR notation. It must be contained by the address space of the virtual network.

	Subnet name	Address range	Addresses
☐	WAF-Subnet	10.0.0.0/24	10.0.0.0 - 10.0.0.255 (256 addresses)
☐	WorkloadSubnet ∨	10.0.1.0/24 ∨	10.0.1.0 - 10.0.1.255 (256 addresses)

Figure 2.102 – Create a virtual network

11. Your **Configure virtual network** section of the **Basics** tab should now look like the following:

Figure 2.103 – Configure virtual network

12. Click **Next : Frontends**.

13. Ensure **Frontend IP address type** is set to **Public**.

14. For **Public IP address**, click **Add new**.

15. In **Add a public IP**, enter a name and click **OK**:

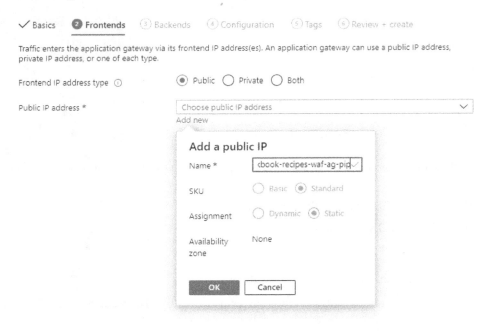

Figure 2.104 – Add a public IP

16. Click **Next : Backends**.

17. Click **Add a backend pool**.

18. In **Add a backend pool**, enter a name, select **Yes** for **Add backend pool without targets**, and then click **Add**:

Figure 2.105 – Add a backend pool

19. Click **Next : Configuration**.

20. Click **Add a routing rule** under **Routing rules**:

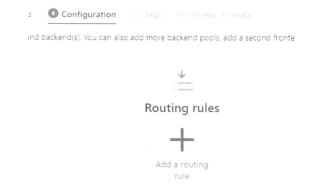

Figure 2.106 – Add a routing rule

21. In the **Add a routing rule** blade, enter a name and set **Priority** as 1:

Figure 2.107 – Configure a routing rule

22. In the **Listener** tab, set the following:

   - Enter a name for **Listener name**

   - Ensure **Frontend IP** is set to **Public**

   - Ensure **Protocol** is set to **HTTP**

   - Ensure **Port** is set to 8 0

   - Ensure **Listener type** is set to **Basic**

   - Ensure **Error page url** is set to **No**

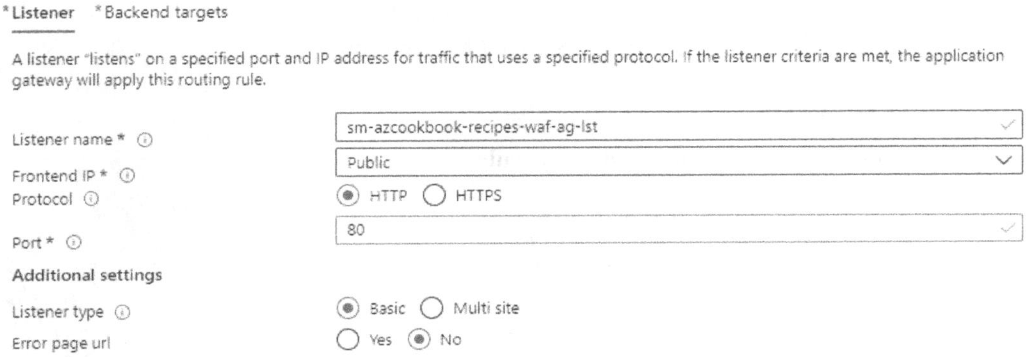

*Listener   *Backend targets

A listener "listens" on a specified port and IP address for traffic that uses a specified protocol. If the listener criteria are met, the application gateway will apply this routing rule.

Listener name * ⓘ	sm-azcookbook-recipes-waf-ag-lst ✓
Frontend IP * ⓘ	Public ∨
Protocol ⓘ	◉ HTTP ○ HTTPS
Port * ⓘ	80 ∨
Additional settings	
Listener type ⓘ	◉ Basic ○ Multi site
Error page url	○ Yes ◉ No

Figure 2.108 – Configure routing rule listener

23. In the **Backend targets** tab, ensure **Backend pool** is set for **Target type**; for **Backend target**, select the backend pool you created earlier in this recipe:

Choose a backend pool to which this routing rule will send traffic. You will also need to specify a set of Backend settings that define the behavior of the routing rule.

Target type	◉ Backend pool ○ Redirection
Backend target * ⓘ	sm-azcookbook-recipes-waf-ag-bep ∨
	Add new

Figure 2.109 – Configure routing rule backend target

24. For **Backend settings**, click **Add new**.

25. In the **Add Backend setting** blade, enter a name for **Backend settings name**; leave all other settings at the defaults, and then click **Add**:

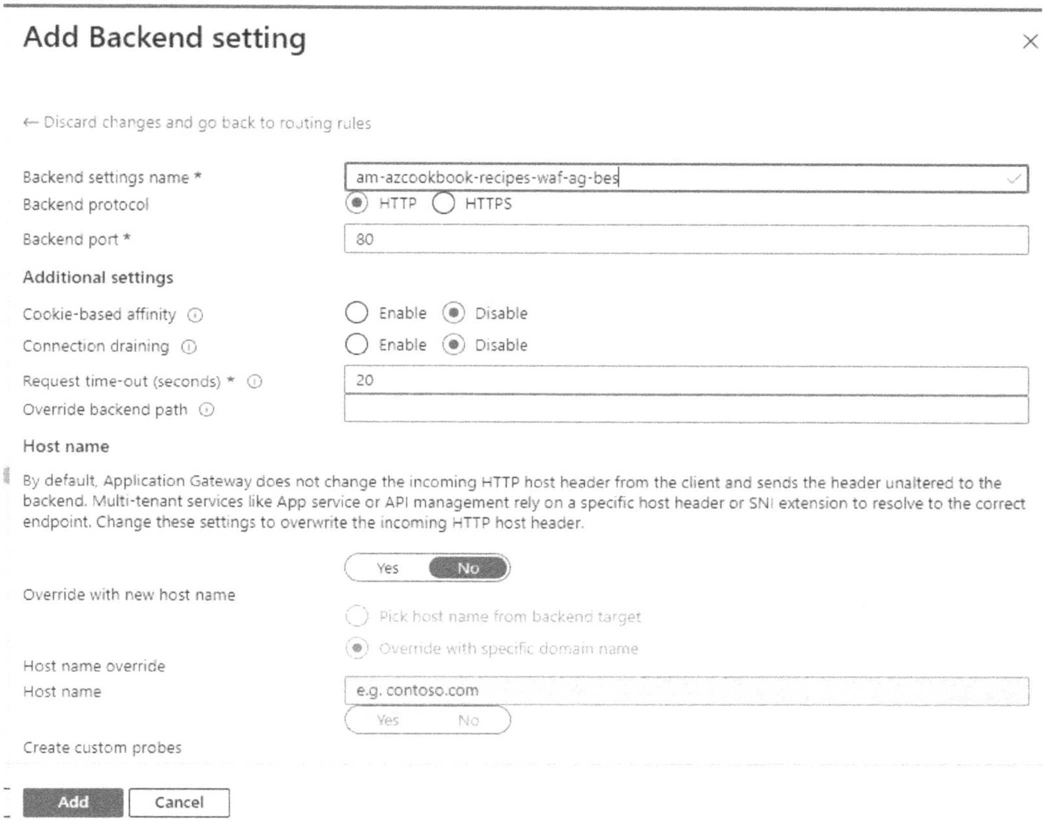

Add Backend setting   ✕

← Discard changes and go back to routing rules

Backend settings name *	am-azcookbook-recipes-waf-ag-bes ✓
Backend protocol	⦿ HTTP ◯ HTTPS
Backend port *	80

Additional settings

Cookie-based affinity ⓘ	◯ Enable ⦿ Disable
Connection draining ⓘ	◯ Enable ⦿ Disable
Request time-out (seconds) * ⓘ	20
Override backend path ⓘ	

Host name

By default, Application Gateway does not change the incoming HTTP host header from the client and sends the header unaltered to the backend. Multi-tenant services like App service or API management rely on a specific host header or SNI extension to resolve to the correct endpoint. Change these settings to overwrite the incoming HTTP host header.

Override with new host name	Yes **No**
	◯ Pick host name from backend target
	⦿ Override with specific domain name
Host name override	
Host name	e.g. contoso.com
Create custom probes	Yes No

Add   Cancel

Figure 2.110 – Add routing rule backend setting

26. Once you are returned to the **Add a routing rule** screen, click **Add**:

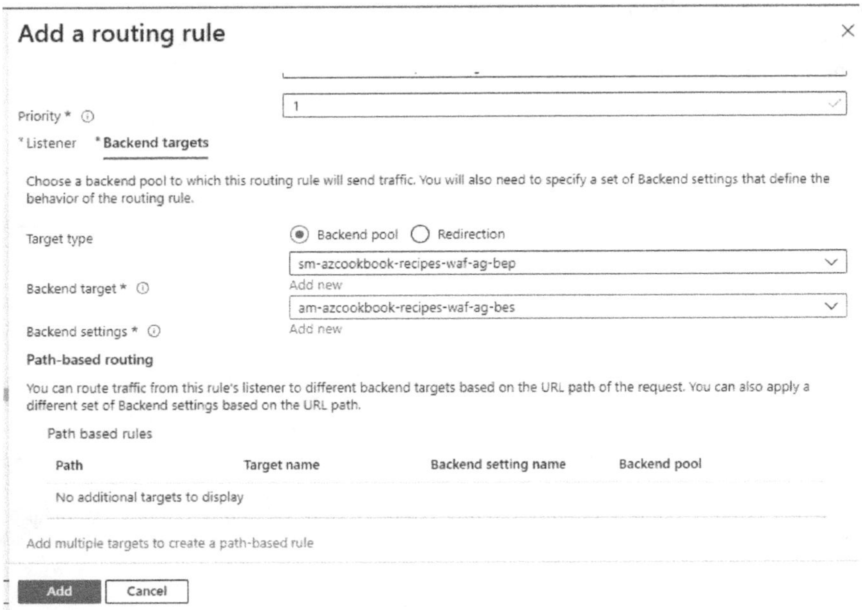

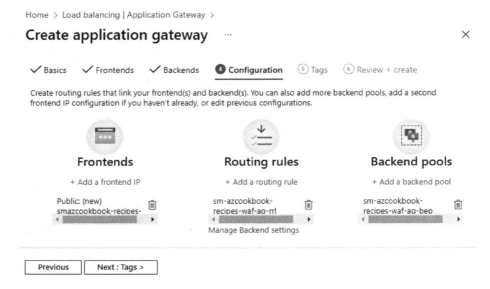

Figure 2.111 – Add a routing rule

27. Once you have returned to the **Configuration** tab, click **Next : Tags**:

Figure 2.112 – Configuration screen

28. Enter any tags if required, then click **Next : Review + create**.

29. Click **Create** on the **Review + create** tab:

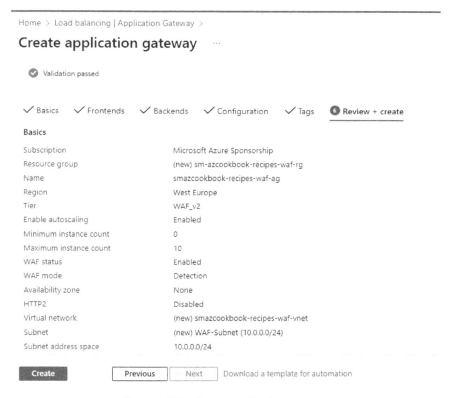

Figure 2.113 – Create application gateway

30. A notification will display that the resource deployment succeeded:

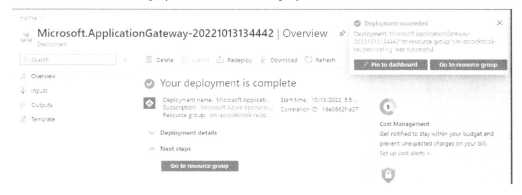

Figure 2.114 – Deployment succeeded

The task to create an application gateway with a WAF enabled is complete. We add our workload server to a backend pool in the next task.

Task – adding a server to the backend pool

Perform the following steps:

1. Navigate to your application gateway in the Azure portal, click on **Backend pools** under the **Settings** section, and click on your backend pool:

Figure 2.115 – Configuring application gateway

2. From the **Edit backend pool** blade, under **Backend targets**, for **Target type**, select **Virtual machine**, and for **Target**, select the virtual machine we created as part of this recipe. Click **Save**:

Figure 2.116 – Edit backend pool

3. You will receive a notification that the deployment succeeded:

Figure 2.117 – Deployment succeeded

The task to create a workload server virtual machine with IIS installed is complete. In the next task, we test access to the web server through the WAF-enabled application gateway.

Task – testing the application gateway with WAF enabled

Perform the following steps:

1. From your application gateway, find your public IP address:

Figure 2.118 – Application gateway public IP address

2. Copy and paste the public IP address into a browser, and ensure you receive a page served by your virtual machine using IIS. This confirms the successful configuration of the application gateway and your enabled WAF:

Figure 2.119 – Successful web server page access

The task to test access to the web application is complete. In the next task, we clean up the resources created in this recipe.

Task – cleaning up resources

Perform the following steps:

1. In the search box in the Azure portal, type `resource groups` and select **Resource Groups** from the listed **Services** results.

2. On the **Resource groups** page, select the resource group we created for this recipe and click **Delete resource group**; this will delete all the resources created as part of this recipe.

The task to clean up the resources created in this recipe is complete.

How it works...

We looked at implementing a WAF as an integrated Azure Application Gateway service component in this recipe. We created a virtual machine with IIS installed as a workload server to test application access.

The WAF uses the **Open Web Application Security Project (OWASP) ModSecurity (ModSec)** core rule for application protection. This provides application protection against the *OWASP Top 10 vulnerabilities*, such as **cross-site scripting** and **injection attacks**; an injection attack example is **SQL injection**.

Further information can be found at the following URL: `https://owasp.org/www-project-modsecurity-core-rule-set`.

The Azure WAF deployment topology, when enabled as a component of the Azure Application Gateway service, is represented in the following diagram:

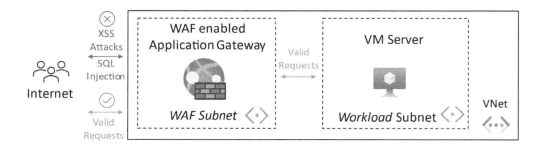

Figure 2.120 – Azure WAF deployment topology

There's more...

WAF protection can also be implemented using the **Azure Front Door** service to provide centralized cross-region protection for your web applications from vulnerabilities.

We can also protect web application resources in other cloud environments or on-premises networks using the WAF as the entry point to access these applications.

See also

Should you require further information, you can refer to the following Microsoft Learn articles:

* *Azure Application Gateway documentation*: `https://learn.microsoft.com/en-us/azure/application-gateway`

* *Web Application Firewall documentation*: `https://learn.microsoft.com/en-us/azure/web-application-firewall`

* *Web Application Firewall (WAF) on Azure Front Door*: `https://learn.microsoft.com/en-us/azure/frontdoor/web-application-firewall`

* *Tutorial: Create a Web Application Firewall policy on Azure Front Door using the Azure portal*: `https://learn.microsoft.com/en-us/azure/web-application-firewall/afds/waf-front-door-create-portal`

Implementing Azure DDoS

In the previous section on implementing a WAF-enabled application gateway, we looked at protecting our web applications that are vulnerable to **Layer 7** inbound web protocol attacks.

We continue, in this section, with our *defense-in-depth* strategy and look at additional protection methods for the protection of the network.

We will look at protecting **Layers 3** and **4** of our network against **Distributed Denial of Service (DDoS)** attacks using the **Azure DDoS Protection Standard service**.

This recipe will teach you how to implement an **Azure DDoS protection plan** to protect your Azure virtual network(s).

We will take you through creating a DDoS protection plan and enabling protection for new and existing virtual networks, and provide information on how you may perform validation testing using Microsoft-supported third-party tools.

Getting ready

This recipe requires the following:

- A device with a browser, such as Edge or Chrome, to access the Azure portal: `https://portal.azure.com`

- You should sign in with an account that has the **Owner** or **Contributor** role for the **Azure subscription**

> **Pricing caution!**
>
> You should be aware that the **DDoS Protection service** has a *fixed monthly* price.
>
> You will be charged the monthly fee *regardless of usage* if the service is active for the whole month. However, you will be prorated if only used for a portion of the month.
>
> This is important to consider for this recipe as the service is **$2,944** for an active month at the time of writing. We have included steps for removing this resource so that you do not receive a bill for longer than it was active in your environment.
>
> *Author disclaimer:* It is **strongly recommended** that you *do not* leave this service running for any longer than required to implement and test this service as part of this recipe in a testing and evaluation scenario, due to the cost implications.
>
> Further information on pricing can be found at this URL: `https://azure.microsoft.com/en-gb/pricing/details/ddos-protection`.

How to do it...

This recipe consists of the following tasks:

- Create a DDoS protection plan.
- Enable DDoS protection for a new virtual network.
- Enable DDoS protection for an existing virtual network.
- View protected resources.
- Clean up resources.

Task – creating a DDoS protection plan

Perform the following steps:

1. From the top left of the Azure portal, click **Create a resource**:

Figure 2.121 – Create a resource

2. Type DDoS in the search box and select **DDoS protection plan** from the results:

Figure 2.122 – Search for DDoS

3. Select **Create** on the **DDoS protection plan** page:

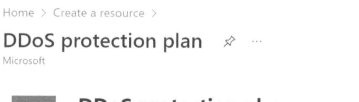

Figure 2.123 – Create a DDoS protection plan

4. From the **Basics** tab, under the **Project details** section, set **Subscription** as required:

Figure 2.124 – Set Subscription

5. For **Resource group**, click **Create new**.

6. Enter a name and click **OK**:

Figure 2.125 – Create a resource group

7. Type in a name and select a region as required:

Figure 2.126 – Set instance details

8. Click **Review + create**.

9. Review the important pricing information and then click **Create**:

Figure 2.127 – Create a plan

The task to create a DDoS protection plan is complete. We will enable DDoS protection for a new virtual network in the next task.

Task – enabling DDoS protection for a new virtual network

Perform the following steps:

1. In the search box in the Azure portal, type `virtual network` and select **Virtual networks** from the listed **Services** results:

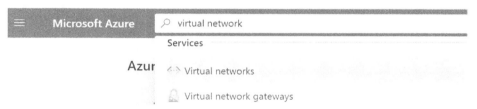

Figure 2.128 – Search for virtual networks

2. On the **Virtual networks** screen, click **+ Create**:

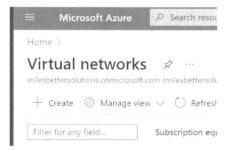

Figure 2.129 – Create a virtual network

3. From the **Basics** tab, set **Subscription** and **Resource group** to the same values we set in the previous task in this recipe to create the virtual machine.

4. Type a name and region as required, then click **Next : IP Addresses**.

5. Leave the settings as the default on the **IP Addresses** tab and click **Next : Security**.

6. From the **Security** tab, click **Enable**, and also read the information pop-up screen by hovering the mouse over the **i** information symbol:

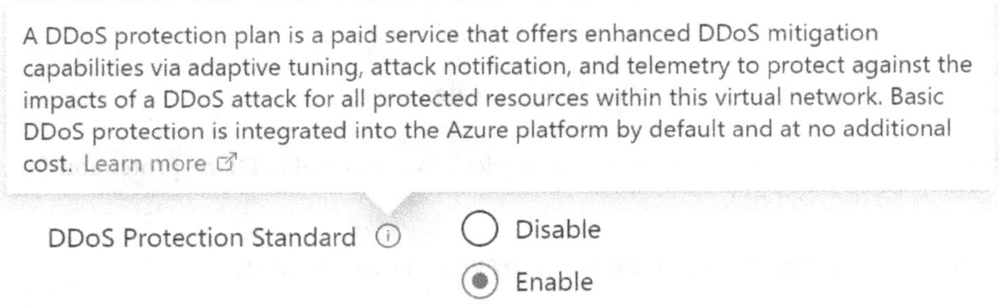

> A DDoS protection plan is a paid service that offers enhanced DDoS mitigation capabilities via adaptive tuning, attack notification, and telemetry to protect against the impacts of a DDoS attack for all protected resources within this virtual network. Basic DDoS protection is integrated into the Azure platform by default and at no additional cost. Learn more ☐

DDoS Protection Standard ⓘ   ◯ Disable

                                               ⦿ Enable

Figure 2.130 – Enable protection plan

7. Select the DDoS protection plan you created in the previous task:

DDoS protection plan *

Choose a DDoS protection plan.   ⌄

Firewall ⓘ

sm-azcookbook-recipes-ddos-ProtectionPlan

Figure 2.131 – Select the protection plan

8. Click **Review + create**, then click **Create** once the validation has passed.

The task to set DDoS protection for a new virtual network is complete. In the next task, we enable DDoS protection for an existing virtual network.

Task – enabling DDoS protection for an existing virtual network

Perform the following steps:

1. If you do not have an existing virtual network without DDoS protection enabled, create a new virtual network to use this recipe as a *getting ready* task.

2. Within the Azure portal, navigate to an existing virtual network on which you wish to enable DDoS protection.

3. On the **Virtual network** screen, select **DDoS protection** under the **Settings** section.

4. In the **DDoS protection** blade for the virtual network, select **Enable** and select the DDoS protection plan to use for this virtual network, then click **Save**

Figure 2.132 – Enable a DDoS protection plan

The task to enable DDoS protection for an existing virtual network is complete. In the next task, we view the protected resources in the DDoS protection plan.

Task – view protected resources

Perform the following steps:

1. Navigate to the **DDoS protection plans** screen; you can use the *search box* or select **All services** and filter to locate **DDoS protection plans**:

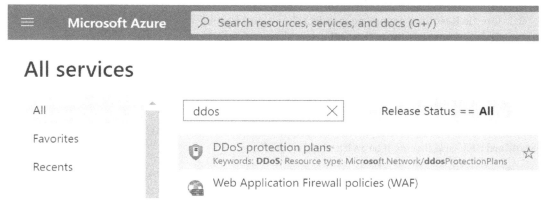

Figure 2.133 – Search for DDoS protection plans

2.   On the **DDoS protection plans** screen, click on your DDoS protection plan:

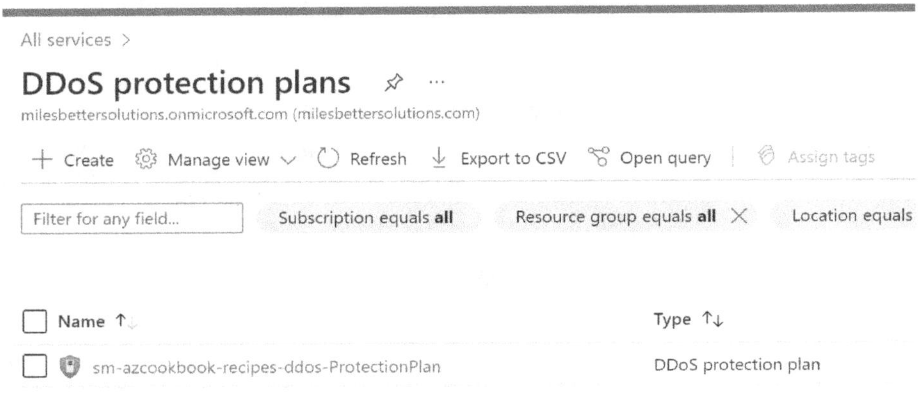

All services >

DDoS protection plans 📌 ...
milesbettersolutions.onmicrosoft.com (milesbettersolutions.com)

+ Create   ⚙ Manage view ∨   ↻ Refresh   ↓ Export to CSV   ⬚ Open query | ⊘ Assign tags

| Filter for any field... | Subscription equals **all** | Resource group equals **all** ✕ | Location equals |

☐ Name ↑	Type ↑↓
☐ 🛡 sm-azcookbook-recipes-ddos-ProtectionPlan	DDoS protection plan

Figure 2.134 – DDoS protection plan

3.   Click on **Protected resources** under the **Settings** section, and you will see the virtual networks that we enabled to be protected.

The task to view the protected resources in our protection plan is complete. In the next task, we clean up the resources created in this recipe.

Task – cleaning up resources

Perform the following steps:

1.   In the search box in the Azure portal, type resource groups and select **Resource Groups** from the listed **Services** results.

2.   On the **Resource groups** page, select the resource group we created for this recipe and click **Delete resource group**; this will delete all the resources created as part of this recipe.

The task to clean up the resources created in this recipe is complete.

How it works...

In this recipe, we looked at implementing **Azure DDoS Protection** using the **DDoS Protection Standard SKU** and linking it to a virtual network.

A **DDoS protection plan** is created when **DDoS Protection Standard** is enabled. To get full protection, you can link virtual networks from multiple subscriptions of the same Azure AD tenant.

Azure as a platform has inherent DDoS network protection; however, this is there to protect at the infrastructure level, not at the individual customer-workload level.

By implementing this protection using the cost-based **DDoS Protection Standard SKU** at your specific workload layer, you can ensure targeted protection that is tuned to your web application traffic patterns. This provides much tighter protection than the generic volumetric infrastructure-level protection.

The following table shows a features comparison between the Azure platform-provided *no-cost* **DDoS infrastructure protection** and the *cost-based* **DDOS Protection Standard**:

Feature	DDoS Infrastructure Protection	DDoS Protection Standard SKU
Automatic attack mitigation	Yes	Yes
Active traffic monitoring and always-on detection	Yes	Yes
Application-based mitigation policies	No	Yes
Availability guarantee	No	Yes
Cost protection	No	Yes
DDoS rapid response support	No	Yes
Metrics and alerts	No	Yes
Mitigation flow logs	No	Yes
Mitigation policy customizations	No	Yes
Mitigation reports	No	Yes

Table 2.1 – Azure DDoS Protection capabilities comparison

The public IPs will be protected when associated with virtual machines (including **Network Virtual Appliances** (**NVAs**)), load balancers, application gateways, Azure Firewall, Azure Bastion, and VPN gateways. Customers' own custom IP prefixes brought into Azure are also protected.

The **DDoS Standard Protection** service can mitigate the following types of attacks:

- Volumetric attacks
- Protocol attacks (*Layers 3 and 4*)
- Resource-layer attacks (*Layer 7*)

Figure 2.135 – Azure DDoS Protection

There's more...

It is important to note that should you wish to validate that the **DDoS service** will protect your resources from a DDoS attack, Microsoft will only allow the simulation of attacks using the following penetration testing partners:

- **BreakingPoint Cloud**: `https://www.keysight.com/us/en/products/network-security/breakingpoint-cloud`
- **Red Button**: `https://www.red-button.net`

See also

Should you require further information, you can refer to the following Microsoft Learn articles:

- *Azure DDoS Protection documentation*: `https://learn.microsoft.com/en-us/azure/ddos-protection`
- *Azure DDoS Protection reference architectures*: `https://learn.microsoft.com/en-us/azure/ddos-protection/ddos-protection-reference-architectures`

3

Securing Remote Access

In the previous chapter, we covered recipes that provided security for Azure networks.

Before implementing resources in Azure, we should consider how we provide remote access in a secure, controlled, and auditable manner.

This chapter looks at how we can provide this protection for our networks and the resources they access.

We will follow on from the network security aspects of the previous chapter, breaking down the chapter into recipes to securely extend an on-premises site into Azure through an encrypted virtual network gateway service using **Azure Network Adapter**.

We will look at the **Azure Bastion service**, which allows us RDP and SSH access without needing to open these management ports or provide public IPs for resources on the virtual network.

Finally, we will cover how to minimize exposure for our Azure **Virtual Machines** (**VMs**) by locking down inbound traffic and restricting access to the management ports through **Just-in-Time** (**JIT**) access.

By the end of this chapter, you will have learned about the following aspects of secure remote access:

- Implementing **Azure Network Adapter**
- Implementing the **Azure Bastion service**
- Implementing **JIT VM access**

Technical requirements

For this chapter, it is assumed that you have an Azure AD tenancy and an Azure subscription from completing the recipes in previous chapters of this cookbook. If you skipped straight to this section, the information to create a new Azure AD tenancy and an Azure subscription for these recipes is included in the following list of requirements.

For this chapter, the following are required for the recipes:

- A device with a browser, such as Edge or Chrome, to access the Azure portal at `https://portal.azure.com`.

- An **Azure AD tenancy** and **Azure subscription**; you may use existing ones or sign up for free at `https://azure.microsoft.com/en-us/free`.

- An **Owner** role for the **Azure subscription**.

- An on-premises Windows Server 2019 machine (physical or virtual) with local administrator privilege and internet access. This server should have **Windows Admin Center** (**WAC**) installed and be registered with your Azure subscription; the following Microsoft Learn articles will assist you if you do not have **WAC** installed on a server in your environment or registered with an Azure subscription:

  - `https://learn.microsoft.com/en-us/windows-server/manage/windows-admin-center/overview`
  - `https://learn.microsoft.com/en-us/windows-server/manage/windows-admin-center/azure/azure-integration`

Implementing Azure Network Adapter

As part of a hybrid cloud strategy, it is important to consider how you will implement secure and protected cross-premises connectivity without public IP addressing and exposing vulnerable management ports and protocols to the internet.

This recipe will teach you to implement **Azure Network Adapter** to securely connect your on-premises Window Server to your Azure virtual network using a Point-to-Site encrypted **Virtual Private Network** (**VPN**) connection.

Getting ready

This recipe requires the following:

- A device with a browser, such as Edge or Chrome, to access the Azure portal at `https://portal.azure.com`.

- Access to an Azure subscription, where you have access to the **Owner** role for the **Azure subscription**.

- An on-premises or non-Azure hosted **Windows 2019 server** where we will enable **Azure Network Adapter**; this server will be used to establish the connection to the **Azure target VM**. If this server has any network filtering, such as a firewall, you should allow **RDP** (port `3389`) **outbound**.

- Access to WAC, registered with an Azure subscription, will be used for this recipe. Note that **WAC** does not have to be on the server you are connecting to Azure; it must, though, be able to connect to the server you wish to connect to Azure.

- A Windows Server **Azure VM** to use with this recipe; we will walk through creating this virtual machine as a *Getting ready* task:

  - This will be created *without* an *NSG* attached to its network interface or the VM's subnet.

  - This will *not* have a public IP address associated with its network interface.

  - The virtual network that the VM is connected to will *not* have a VPN gateway service; we will create this as part of the main recipe.

Continue with the following *Getting ready* task for this recipe:

- Creating a virtual machine

Getting ready task – creating a virtual machine

Perform the following steps:

1. Sign in to the Azure portal: `https://portal.azure.com`.

2. From the search box in the Azure portal, type `virtual machines` and select **Virtual machines** from the listed **Services** results.

3. Click **+ Create** from the top-left menu bar on the **Virtual machine** screen and select **Azure virtual machine**.

4. From the **Basics** tab, under the **Project details** section, set the **Subscription** as required.

5. For **Resource group**, click **Create new**.

6. Enter a **Name** value and click **OK**.

7. Under **Instance details**, set the following:

   - **Virtual machine name**: Type a name.

   - **Region**: Select a region.

   - **Availability options**: Select **No infrastructure redundancy required**.

   - **Security type**: Select **Standard**.

   - **Image**: Select **Windows Server 2019 Datacenter – Gen2**.

   - **Size**: Leave the default (or set it as required to reduce recipe costs).

8. Under **Administrator account**, set **Username** and **Password** details as required.

9. Under **Inbound port rules**, set **Public inbound ports** to **None**.

10. Click on **Next : Disks** leave the defaults and then click on **Next : Networking**.

11. Under **Network interface**, leave the defaults for **Virtual network** and **Subnet**.

12. Set **Public IP** to **None** from the dropdown.

13. Set **NIC network security group** to **None**.

14. Check the **Delete NIC when VM is deleted** checkbox.

15. Click **Review + create**.

16. Click **Create** on the **Review + create** tab once validation has passed.

17. A notification will display that the resource deployment succeeded.

The *Getting ready* task for this recipe is complete.

You are now ready to continue the main tasks for this recipe of adding Azure Network Adapter.

How to do it...

This recipe consists of the following tasks:

- Adding Azure Network Adapter to a Windows server
- Connecting to our target Azure VM with RDP

Task – adding Azure Network Adapter

Perform the following steps:

1. Log in to the server where **WAC** is installed and locate the Windows server on which you wish to enable Azure Network Adapter.

2. From **Tools**, navigate or search for Networks, and select **Add Azure Network Adapter**:

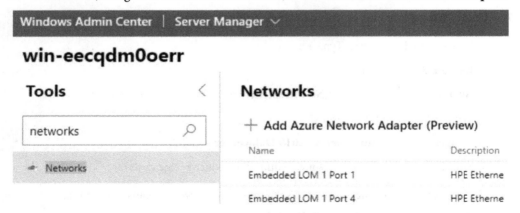

Figure 3.1 – Add Azure Network Adapter

3. From the **Add Azure Network Adapter** blade, enter the following, and then click on **Create**:

- **Subscription**: Set as required.

- **Location**: Set as required.

- **Virtual Network**: Set as required.

- Accept the defaults for **Gateway Subnet**, **Gateway SKU**, **Client Address Space**, and **Authentication Certificate**.

> **Note**
>
> Your **virtual network** should not have an **Azure VPN gateway service** or a **gateway subnet**; if they do, remove these and let the **Add Azure Network Adapter** wizard configure these in this step.

Figure 3.2 – Add Azure Network Adapter settings

4. You will see a confirmation that the creation is taking place and could take up to 35 minutes:

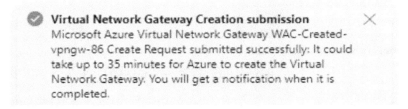

Figure 3.3 – Creation notification

5. You will be notified when the configuration has succeeded:

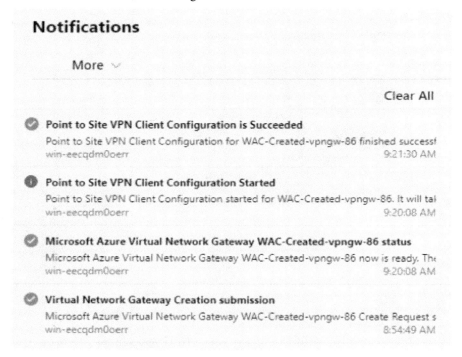

Figure 3.4 – Configuration succeeded notification

6. From the Azure portal, navigate to the **virtual network gateway**, and click on **Point-to-site Sessions** under the **Monitoring** section. You will see the connection from your on-premises server. Note the private IP address:

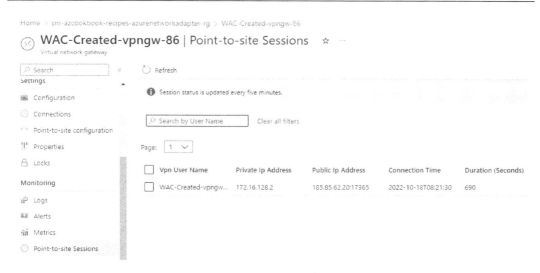

Figure 3.5 – Point-to-site Sessions

This task of creating a VM is complete. In the next task, we will connect to our target server with RDP over the Point-to-Site VPN.

Task – connecting to an Azure Server VM with RDP

Perform the following steps:

1. Navigate to your **target VM** in the Azure portal that you will connect to from your **on-premises server** and note the **private IP address**:

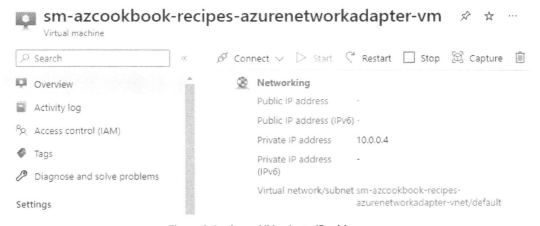

Figure 3.6 – Azure VM private IP address

2. From your on-premises server, open a ,**Command Prompt** (or PowerShell) and enter the `ipconfig` command.

    You will see Azure Network Adapter listed because we enabled it; we can identify it by the private IP address we saw in the **Point-to-site Sessions** blade in the Azure portal. This is represented in the following figure:

```
PPP adapter WACVPN-60277:

   Connection-specific DNS Suffix  . :
   Description . . . . . . . . . . . : WACVPN-60277
   Physical Address. . . . . . . . . :
   DHCP Enabled. . . . . . . . . . . : No
   Autoconfiguration Enabled . . . . : Yes
   IPv4 Address. . . . . . . . . . . : 172.16.128.2(Prefer
   Subnet Mask . . . . . . . . . . . : 255.255.255.255
   Default Gateway . . . . . . . . . :
   NetBIOS over Tcpip. . . . . . . . : Enabled

C:\Users\Administrator>
```

Figure 3.7 – ipconfig command result

3. From your on-premises server, launch an **RDP connection** and enter the private IP address of the Azure VM you noted. When prompted, enter the virtual machine credentials and click **OK**:

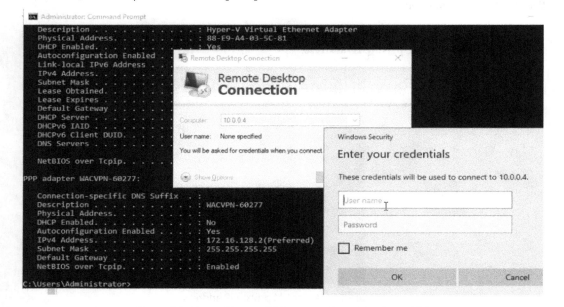

Figure 3.8 – Launching an RDP connection to the Azure VM

4. You will now be successfully logged on to the desktop of your Azure VM:

Figure 3.9 – Successful RDP connection login to the Azure VM

We have completed the task of connecting to our Azure VM via RDP from our on-premises server. In the next task, we will clean up the resources created in this recipe.

Task – cleaning up resources

Perform the following steps:

1. From the search box in the Azure portal, type `resource groups` and select **Resource Groups** from the listed **Services** results.

2. From the **Resource groups** page, select the resource group we created for this recipe, and click **Delete resource group**; this will delete all the resources created as part of this recipe:

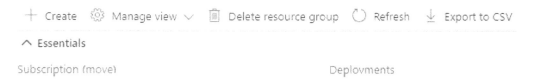

Figure 3.10 – Delete resource group

This task of cleaning up the resources created in this recipe is complete.

How it works...

For this recipe, we looked at implementing Azure Network Adapter for an on-premises Windows server to provide a cross-premises hybrid connectivity solution for our Azure virtual networks.

We set up an Azure VM and a virtual network as our Azure resources to connect to, and then from our on-premises server, we installed WAC, which was registered to our Azure subscription. Once we had confirmed that it was in place, we could configure Azure Network Adapter on our on-premises server, and then we established an RDP connection to an Azure VM using the Point-to-Site VPN created. The following figure represents the topology that we created with this recipe:

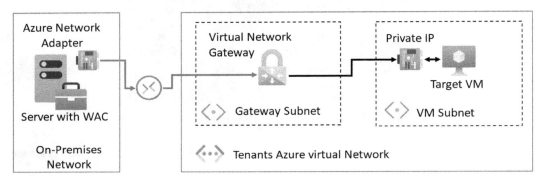

Figure 3.11 – Azure Network Adapter reference topology

Using a Site-to-Site VPN and ExpressRoute circuit are two ways to provide a cross-premises (*east/west traffic*) hybrid connectivity solution, allowing you to connect your on-premises servers to Azure virtual networks. However, both of these can add significant complexity to implementing and operating what may be a very simple use case scenario.

Azure Network Adapter was introduced with Windows Server 2019 as a hybrid connectivity solution that was simple to implement, easy to operate, and cost-effective where a full enterprise-scale connectivity solution was not required.

Through a **one-click solution** using WAC, Azure Network Adapter can provide a solution that connects your on-premises servers to Azure virtual networks using a Point-to-Site VPN.

The following URL provides information on the alternative options for creating a cross-premises hybrid connectivity solution:

- https://learn.microsoft.com/en-us/azure/architecture/reference-architectures/hybrid-networking

There's more...

In this recipe, we simplified the steps to focus on implementing Azure Network Adapter's core steps. We did not use an NSG for the VM subnet or network interface. Regarding network security in a non-lab environment, we should add an NSG to the VM subnet and the network interface for the target VM(s) as appropriate.

In this recipe, as shown in the preceding figure, we installed WAC on the same server where we enabled Azure Network Adapter for simplicity. However, in a real-world scenario, you will have WAC installed on another server on the network.

See also

Should you require further information, you can refer to the following Microsoft Learn article:

- `https://learn.microsoft.com/en-us/windows-server/manage/windows-admin-center/azure/use-azure-network-adapter`

Implementing the Azure Bastion service

Azure Bastion is a Microsoft fully managed RDP/SSH secure remote access connectivity solution for your Azure VM resources. It protects your Azure VMs' vulnerable RDP/SSH management ports without exposing them to the internet or requiring public IP addressing in your virtual network.

Getting ready

This recipe requires the following:

- A device with a browser, such as Edge or Chrome, to access the Azure portal: `https://portal.azure.com`.
- You should sign in with an account that has the **Owner** role for the **Azure subscription**.
- A Windows Server **Azure VM** to use with this recipe; we will step through creating this VM as a *Getting ready* task:
    - To keep this recipe simple, the VM will be created *without* an *NSG* attached to its network interface or the VM's subnet. We will discuss this further in the *There's more* section of this recipe.
    - This will *not* have a public IP address associated with its network interface.
    - The virtual network that the VM is connected to will *not* have a VPN gateway service.

Continue with the following *Getting ready* task for this recipe:

- Creating a virtual machine

Getting ready task – creating a virtual machine

Perform the following steps:

1. Sign in to the Azure portal: `https://portal.azure.com`.

2. From the search box in the Azure portal, type `virtual machines` and select **Virtual machines** from the listed **Services** results.

3. Click **+ Create** from the top-left menu bar on the **Virtual machine** screen and select **Azure virtual machine**.

4. From the **Basics** tab, under the **Project details** section, set the **Subscription** details as required.

5. For **Resource group**, click **Create new**.

6. Enter a **Name** value and click **OK**.

7. Under **Instance details**, set the following:

    -   **Virtual machine name**: Type a name.

    -   **Region**: Select a region.

    -   **Availability options**: Select **No infrastructure redundancy required**.

    -   **Security type**: Select **Standard**.

    -   **Image**: Select **Windows Server 2019 Datacenter – Gen2**.

    -   **Size**: Leave the default (or set it as required to reduce recipe costs).

8. Under **Administrator account**, set **Username** and **Password** details as required.

9. Under **Inbound port rules**, set **Public inbound ports** to **None**.

10. Click **Next : Disks**, leave the defaults, and then click **Next : Networking**.

11. Under **Network interface**, leave the defaults for **Virtual network** and **Subnet**.

12. Set **Public IP** to **None** from the dropdown.

13. Set **NIC network security group** to **None**

14. Check the **Delete NIC when VM is deleted** checkbox.

15. Click **Review + create**.

16. Click **Create** on the **Review + create** tab once *validation* has passed.

17. A notification will display that the resource deployment succeeded.

This *Getting ready* task to create a VM for this recipe is complete.

You are now ready to continue the main tasks for this recipe of adding the Azure Bastion service.

How to do it...

This recipe consists of the following tasks:

- Creating a Bastion
- Connecting to our target Azure VM with a Bastion

Task – creating a Bastion

Perform the following steps:

1. From the search box in the Azure portal, type `bastion` and select **Bastions** from the listed **Services** results.

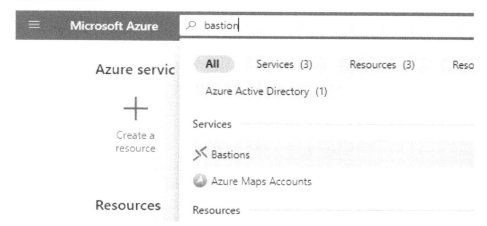

Figure 3.12 – Search for Bastions

2.   On the **Bastions** screen, click **+ Create** or **Create Bastion**:

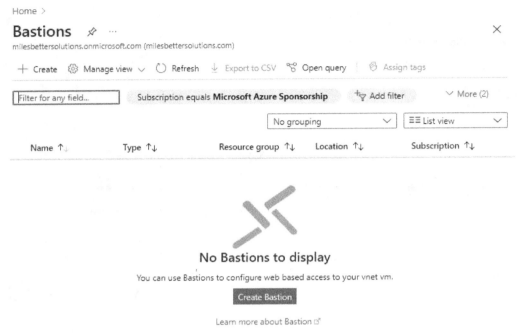

Figure 3.13 – Create a bastion

3.   From the **Basics** tab, under the **Project details** section, set the **Subscription** details as required.

4.   For **Resource group**, click **Create new**.

5.   Enter a **Name** value and click **OK**:

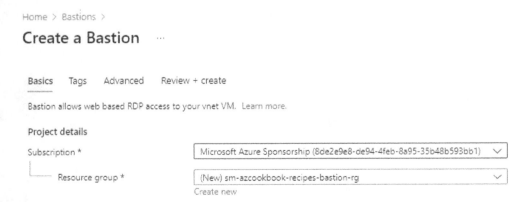

Figure 3.14 – Setting the project details

6. Set the following information under the **Instance details** section:

- **Name**: Enter as required.

- **Region**: Enter the same region selected for our workload VM we created.

- **Tier**: Select **Standard**.

- **Instance count**: Leave the default:

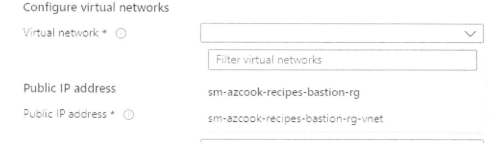

Instance details

Name *	sm-azcookbook-recipes-bastion
Region *	UK South
Tier * ○	Standard
Instance count * ○	2

Figure 3.15 – Setting the instance details

7. From the **Configure virtual networks** section, from **Virtual network**, select the virtual network we created for the VM in the *Getting ready* tasks. If you cannot see your network, ensure you select the same region:

Configure virtual networks

Virtual network * ○	
	Filter virtual networks
Public IP address	sm-azcook-recipes-bastion-rg
Public IP address * ○	sm-azcook-recipes-bastion-rg-vnet

Figure 3.16 – Configure virtual networks

8. Note the message about the requirement for a subnet for a Bastion instance; click **Manage subnet configuration** under **Subnet**:

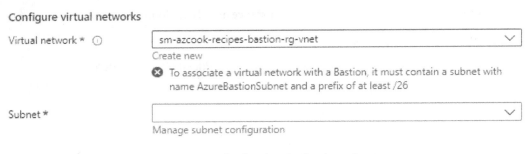

Figure 3.17 – Configuring the Bastion subnet

9.   From the **Subnets** screen, click + **Subnet**:

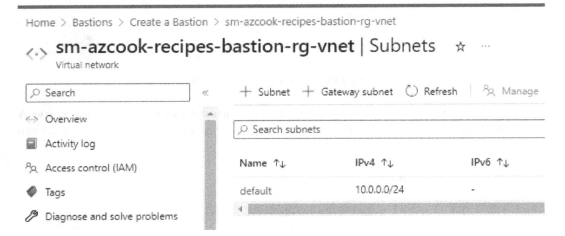

Figure 3.18 – Adding a Bastion subnet

10.   From the **Add subnet** blade, enter AzureBastionSubnet; you must use this exact name:

Figure 3.19 – Naming the Bastion subnet

11. Leave all other settings as their defaults and click **Save**:

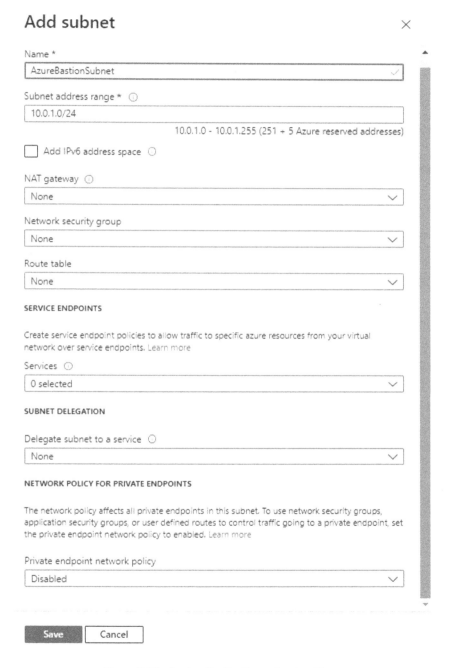

Figure 3.20 – Saving the Bastion subnet settings

12. You will see a notification that the subnet has been created:

Figure 3.21 – Bastion subnet added

13. Close the **Subnets** blade using **X** in the top-right corner:

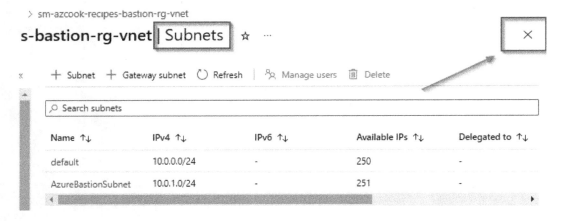

Figure 3.22 – Closing the Subnets blade

14. Leave the settings under the **Public IP address** section as their defaults and click **Review + create**:

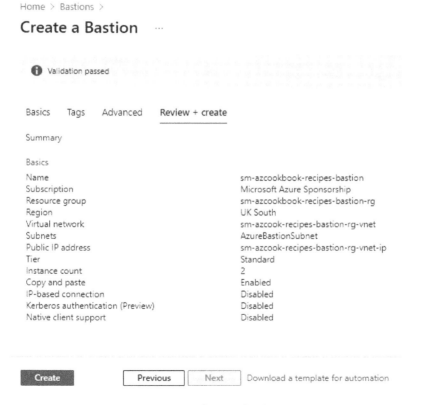

Figure 3.23 – Set public IP address settings

15. Click **Create** on the **Review + create** tab:

Home > Bastions >

Create a Bastion ...

🛈 Validation passed

Basics   Tags   Advanced   Review + create

Summary

Basics

Name	sm-azcookbook-recipes-bastion
Subscription	Microsoft Azure Sponsorship
Resource group	sm-azcookbook-recipes-bastion-rg
Region	UK South
Virtual network	sm-azcook-recipes-bastion-rg-vnet
Subnets	AzureBastionSubnet
Public IP address	sm-azcook-recipes-bastion-rg-vnet-ip
Tier	Standard
Instance count	2
Copy and paste	Enabled
IP-based connection	Disabled
Kerberos authentication (Preview)	Disabled
Native client support	Disabled

Create   Previous   Next   Download a template for automation

Figure 3.24 – Create a Bastion

16. A notification will display that the resource deployment succeeded:

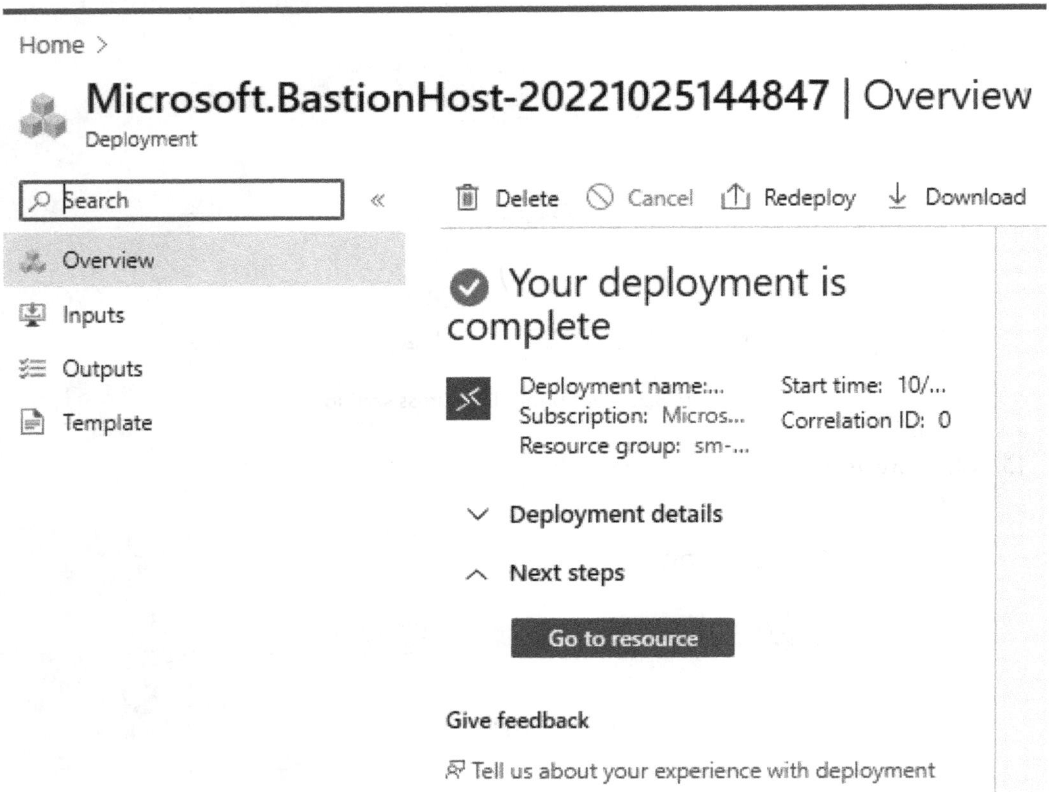

Home >

Microsoft.BastionHost-20221025144847 | Overview
Deployment

Search	«
Overview
Inputs
Outputs
Template

🗑 Delete   ⊘ Cancel   ⬆ Redeploy   ⬇ Download

✅ **Your deployment is complete**

Deployment name:...   Start time: 10/...
Subscription: Micros...   Correlation ID: 0
Resource group: sm-...

∨ **Deployment details**

∧ **Next steps**

[Go to resource]

Give feedback

⤢ Tell us about your experience with deployment

Figure 3.25 – Deployment succeeded

17. Click **Go to resource** and review the settings on the Bastion screen you have created:

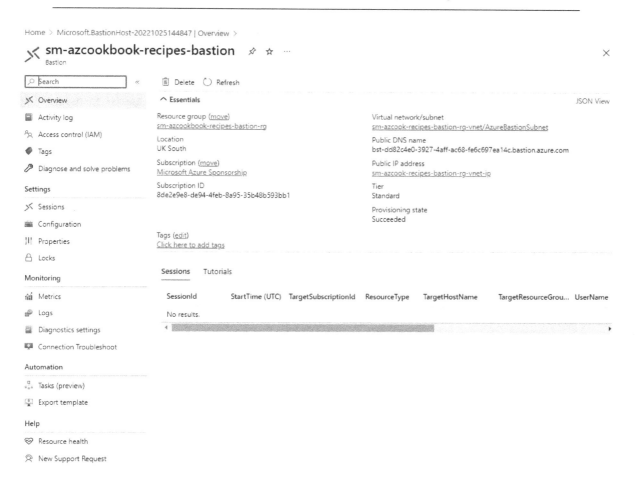

Figure 3.26 – The Bastion screen

This task of creating a Bastion resource is complete. In the next task, we will connect to our target VM with our Bastion.

Task – connecting to our target Azure VM with a Bastion resource

Perform the following steps:

1. Log in to https://portal.azure.com and navigate to the VM to connect via Bastion; ensure it has the **Running** status.

2. From the VM **Overview** page, click on **Connect**, and from the dropdown, select **Bastion**:

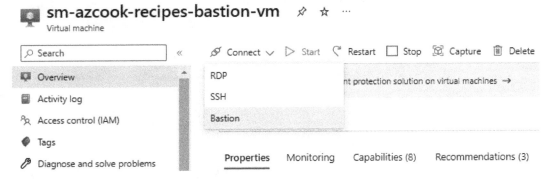

Figure 3.27 – Connecting the VM

3. From the **Bastion** page, click **Connection settings** and make any changes as required, or leave them as their defaults:

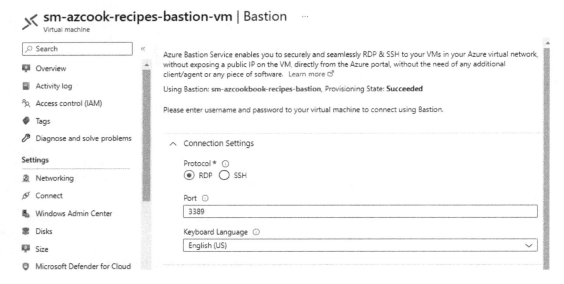

Figure 3.28 – Bastion connection settings

4. Enter the **Username** and **Password** details as required for the VM in question, and then click **Connect**:

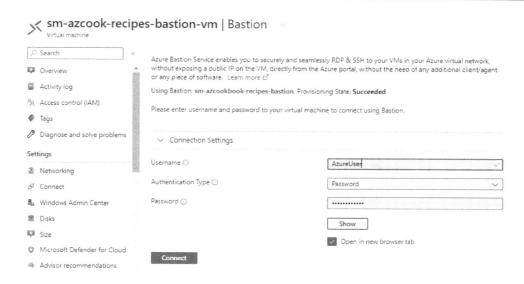

Figure 3.29 – Bastion authentication settings

5. You will be logged onto the VM through the browser using port 443:

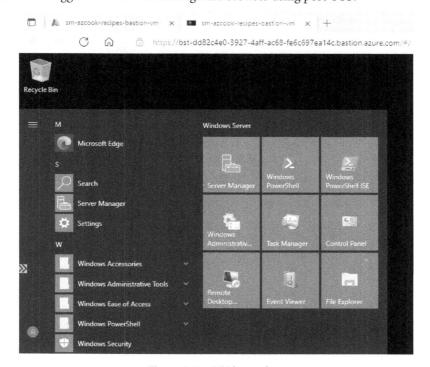

Figure 3.30 – VM logged on

This task of connecting to our Azure VM with a Bastion resource is complete. In the next task, we clean up the resources created in this recipe.

Task – cleaning up resources

Perform the following steps:

1. From the search box in the Azure portal, type `resource groups` and select **Resource Groups** from the listed **Services** results.

2. From the **Resource groups** page, select the resource group we created for this recipe, and click on **Delete resource group**; this will delete all the resources created as part of this recipe:

Figure 3.31 – Delete resource group

This task of cleaning up the resources created in this recipe is complete.

How it works...

For this recipe, we looked at implementing the Azure Bastion service, which provides a remote access solution fully managed by Microsoft that protects the vulnerable RDP/SSH management ports of your Azure VMs.

We set up an Azure VM and a virtual network as our Azure resources that needed protection for remote access. We then deployed Azure Bastion into the virtual network containing the VM we wished to connect securely. Azure Bastion provides a TLS-secured connection and only supports traffic through TCP port 443 without implementing public IP addresses in our virtual network or exposing our vulnerable management ports to the internet.

Azure Bastion provides the remote access connectivity implementation process as follows:

1. Deploy Azure Bastion into an Azure virtual network.

2. Only TCP port 443 communication is allowed to Azure Bastion from the internet; all communication is TLS protected.

3. No public IP addressing is required for the Azure virtual network, and no VM management ports are required to be open or exposed to the internet, significantly improving our security posture and minimizing our threat surface area.

The following illustration represents the Azure Bastion implementation topology:

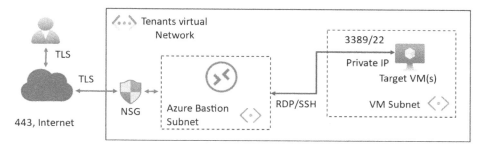

Figure 3.32 – Azure Bastion reference topology

For this recipe, we used an account that had the Owner or Contributor role for the Azure subscription.

The subnet used for Azure Bastion must be named **AzureBastionSubnet**; the minimum size is **/26** or larger – for example, **/24**.

There's more...

In this recipe, we simplified the steps to focus on the core steps of implementing the Azure Bastion service. This meant we did not use an NSG for the VM subnet or network interface, and we carried out the recipe and accessed the VM via the Bastion service with a user that had an Owner role for the subscription. This section will address additional information related to both aspects: the required roles to access a VM.

To access a VM, you will need the following roles, working based on the least privileges:

- The Reader role on the VM resource

- The Reader role on the network interface with the private IP address of the VM

- The Reader role on the virtual network of the target VM(s)

- The Reader role on the Azure Bastion resource

In our recipe, both roles were assigned and inherited from the Owner role for our subscription; as mentioned, working based on the principle of least privilege, you should assign roles to users with the lowest level of access required to perform their tasks in production environments.

Regarding network security, as appropriate, we should add an NSG to the VM subnet or the network interface for the target VM(s).

The following figure represents an NSG added to the VM subnet:

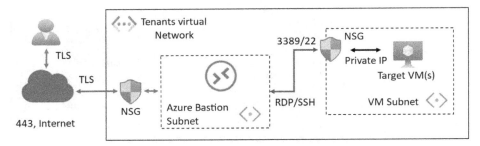

Figure 3.33 – Azure Bastion reference topology with an NSG for target VM subnet

When establishing the RDP connection with a target VM with an NSG associated with the VM subnet or VM network interface, you must ensure that inbound rules are created for the RDP protocol – port 3389 for Windows VMs and the SSH protocol and port 22 for Linux VMs.

See also

Should you require further information, you can refer to the following Microsoft Learn articles:

- https://azure.microsoft.com/en-us/products/azure-bastion/#overview
- https://learn.microsoft.com/en-us/training/modules/intro-to-azure-bastion
- https://learn.microsoft.com/en-us/training/modules/connect-vm-with-azure-bastion
- https://learn.microsoft.com/en-us/azure/bastion
- https://learn.microsoft.com/en-us/azure/bastion/tutorial-create-host-portal

Implementing JIT VM access

As we continue our defense-in-depth journey, in this section, we will look at another solution to increase your security posture, reduce exposure to attacks, and disrupt and block the path of threat actors.

This recipe looks at limiting access to vulnerable management ports on a VM using the **JIT VM** feature.

We will cover how to enable JIT access for VMs from within the **Virtual machines** blade using the Azure portal, requesting access for a JIT-enabled VM, and activity auditing.

Getting ready

This recipe requires the following:

- A device with a browser, such as Edge or Chrome, to access the Azure portal at `https://portal.azure.com`

- You should sign in with an account that has the **Owner** role for the **Azure subscription**

- **Microsoft Defender for Servers Plan 2** must be enabled on the subscription used for this exercise; we will step through this process in the following *Getting ready* tasks if not already enabled

- An **Azure VM**; we will walk through this process in the following *Getting ready* tasks

Continue with the following *Getting ready* task for this recipe:

- Creating a VM

Getting ready task – creating a VM

Perform the following steps:

1. Sign in to the Azure portal at `https://portal.azure.com`.
2. From the search box in the Azure portal, type `virtual machines` and select **Virtual machines** from the listed **Services** results.
3. Click **+ Create** from the top-left menu bar on the **Virtual machine** screen and select **Azure virtual machine**.
4. From the **Basics** tab, under the **Project details** section, set the **Subscription** details as required.
5. For **Resource group**, click **Create new**.
6. Enter a **Name** value and click **OK**.
7. Under **Instance details**, set the following:

   - **Virtual machine name**: Type a name.
   - **Region**: Select a region.
   - **Availability options**: Select **No infrastructure redundancy required**.
   - **Security type**: Select **Standard**.
   - **Image**: Select **Windows Server 2019 Datacenter – Gen2**.
   - **Size**: Leave the default (or set it as required to reduce recipe costs).

8. Under **Administrator account**, set the **Username** and **Password** details as required.
9. Under **Inbound port rules**, set **Public inbound ports** to **Allow selected ports**.

10. For **Select inbound ports**, check **RDP (3389)**:

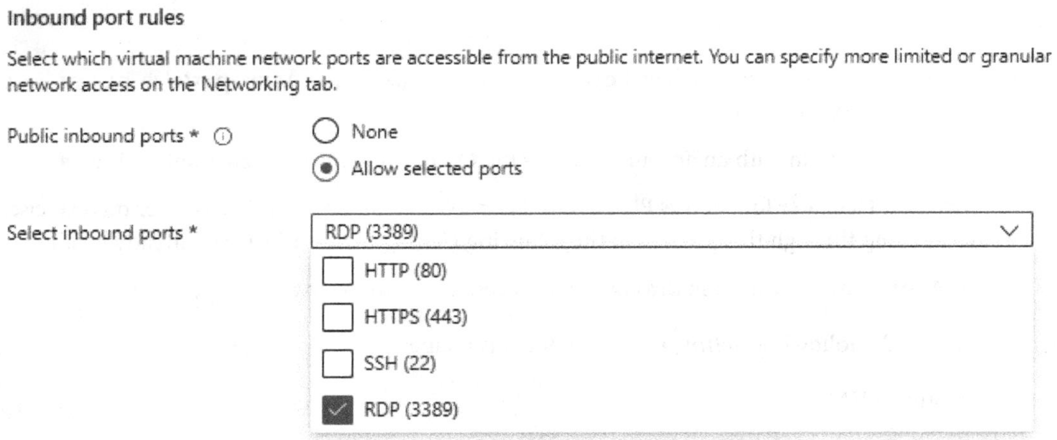

Inbound port rules

Select which virtual machine network ports are accessible from the public internet. You can specify more limited or granular network access on the Networking tab.

Public inbound ports * ⓘ
- ◯ None
- ◉ Allow selected ports

Select inbound ports *

RDP (3389)	⌄
☐ HTTP (80)	
☐ HTTPS (443)	
☐ SSH (22)	
☑ RDP (3389)	

Figure 3.34 – Inbound port rules

11. Click **Next : Disks**, leave the defaults, and then click **Next : Networking**.

12. Under **Network interface**, leave the defaults for **Virtual network** and **Subnet**.

13. Leave the default for **Public IP**.

14. Ensure **NIC network security group** is set to **Basic**.

15. Ensure the **RDP (3389)** inbound port is selected.

16. Check the **Delete public IP and NIC when VM is deleted** checkbox.

17. Click **Review + create**.

18. Click **Create** on the **Review + create** tab once *validation* has passed.

19. A notification will display that the resource deployment succeeded

The *Getting ready* task for this recipe is complete.

You are now ready to continue the main tasks for this recipe of implementing JIT access for a VM.

How to do it...

This recipe consists of the following tasks:

- Enabling JIT access for a VM
- Requesting JIT-enabled VM access

Task – enabling JIT access for a VM

Perform the following steps:

1. Log in to the portal at `https://portal.azure.com` and navigate to the VM to connect via Bastion.

2. Click **Configuration** from your **Virtual machine** page under the **Settings** section of the left-hand menu:

Figure 3.35 – VM configuration

3. Click on **Upgrade your Microsoft Defender for Cloud subscription to enable a just-in-time access**:

Figure 3.36 – Upgrading your Defender for Cloud subscription

4. For this next step, we will enable a 30-day free trial of **Defender for Cloud**. Ensure the subscription for which to enable Defender for Cloud is checked, and click on **Upgrade**:

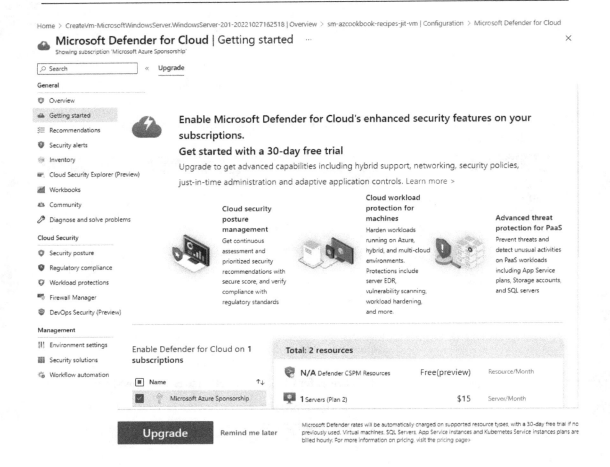

Figure 3.37 – Enabling your free trial

5.   You will receive a notification that your trial has started:

Figure 3.38 – Trial started notification

6.   Close the **Microsoft Defender for Cloud** page to return to the virtual machine configuration page.

7. From the virtual machine configuration page, click on the **Enable just-in-time** button; you may need to refresh the browser to reload the page to view:

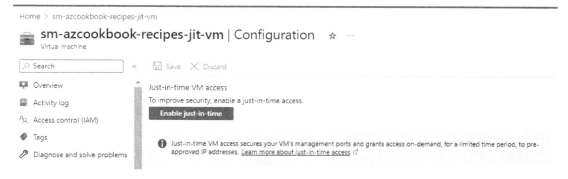

Figure 3.39 – Enabling access

8. You will receive a notification that access was enabled:

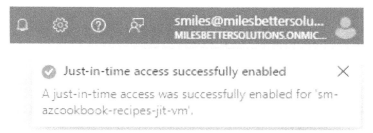

Figure 3.40 – Access enabled notification

9. Click **Open Microsoft Defender for Cloud**:

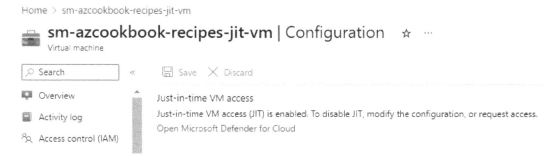

Figure 3.41 – Opening Defender for Cloud

10. You will now see the **Just-in-time VM access** page, showing your VMs under the **Configured** tab:

Home > sm-azcookbook-recipes-jit-vm | Configuration >

Just-in-time VM access 📌 ⋯   ✕
Last week

> **What is just-in-time VM access?**
> **How does it work?**

Virtual machines

Configured   Not Configured   Unsupported

VMs for which the just-in-time VM access control is already in place. Presented data is for the last week.

1 VMs   Request access

🔍 Search to filter items...

	Virtual machine ↑↓	Approved ↑↓	Last access ↑↓	Connection details	Last user ↑↓	
☐ 🖥	sm-azcookbook-recipes...	0 Requests	N/A	🛡 -	N/A	⋯

Figure 3.42 – Access screen

11. You can view and edit the settings by clicking on the three dots at the end of the VM row. Click on **Edit**:

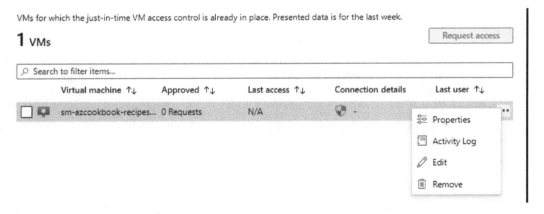

VMs for which the just-in-time VM access control is already in place. Presented data is for the last week.

1 VMs   Request access

🔍 Search to filter items...

	Virtual machine ↑↓	Approved ↑↓	Last access ↑↓	Connection details	Last user ↑↓	
☐ 🖥	sm-azcookbook-recipes...	0 Requests	N/A	🛡 -		⋯

🔲 Properties
🗔 Activity Log
✏ Edit
🗑 Remove

Figure 3.43 – Editing the configuration

12. From the **JIT VM access configuration** screen, you view the default configuration settings:

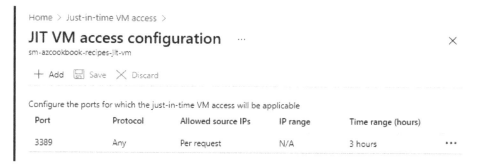

Figure 3.44 – Access configuration

13. Click on the configuration line item, which will open the **Add port configuration** page. Set it as required and then click **OK**:

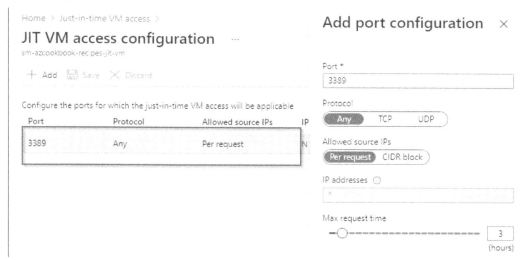

Figure 3.45 – Editing access configuration

This task of enabling JIT access for a VM is complete. For the next task, we will request access to a JIT-enabled VM.

Task – requesting JIT-enabled VM access

Perform the following steps:

1. Log in to https://portal.azure.com and navigate to the VM to access; ensure it has the **Running** status.

2. From the VM **Overview** page, click **Connect**, and from the dropdown, select **RDP**.

3. From the **Connect** page, click on the **Request access** button:

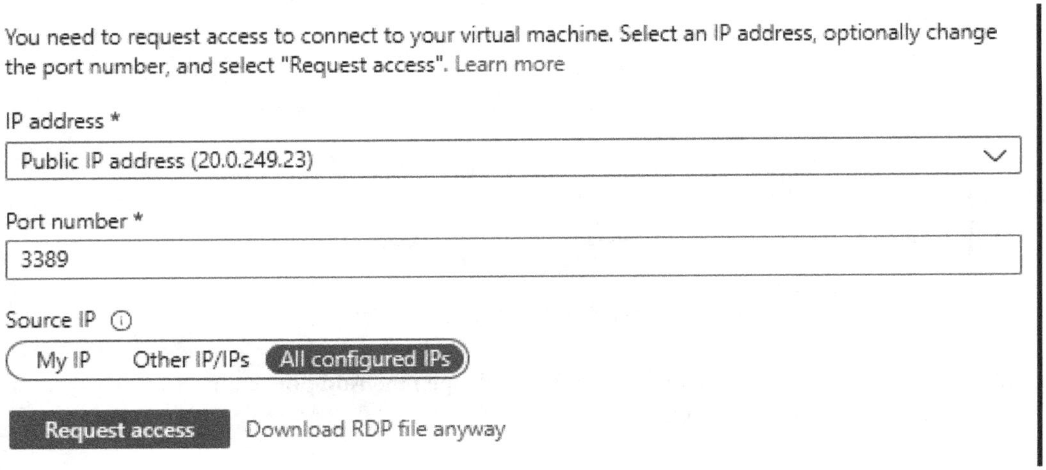

You need to request access to connect to your virtual machine. Select an IP address, optionally change the port number, and select "Request access". Learn more

IP address *

Public IP address (20.0.249.23)   ⌄

Port number *

3389

Source IP ⓘ

(My IP   Other IP/IPs   All configured IPs)

Request access   Download RDP file anyway

Figure 3.46 – VM connection page

4. You will receive a notification that the access request has been approved:

Figure 3.47 – Request approved notification

5. You will now see a message that access has been approved, and you will be able to download the RDP file for access to the VM:

Figure 3.48 – Access approved

6. From the **Just-in-time VM access** page, under the **Configured** tab, click on the three dots at the end of the VM row, click on **Activity Log**, and view the listed operations; you may also click **Download as CSV**.

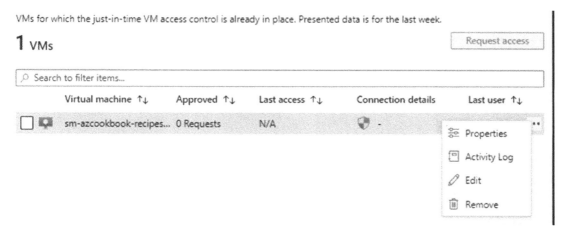

Figure 3.49 – Auditing access

This task of enabling JIT access for a VM is complete. In the next task, we will clean up the resources created in this recipe.

Task – cleaning up resources

Perform the following steps:

1. From the search box in the Azure portal, type `resource groups` and select **Resource Groups** from the listed **Services** results.

2. From the **Resource groups** page, select the resource group we created for this recipe, and click on **Delete resource group**; this will delete all the resources created as part of this recipe:

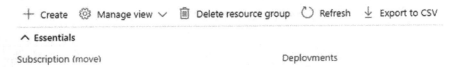

Figure 3.50 – Delete resource group

3. Navigate to Defender for Cloud in the Azure portal, and click on **Environment settings** under the **Management** section:

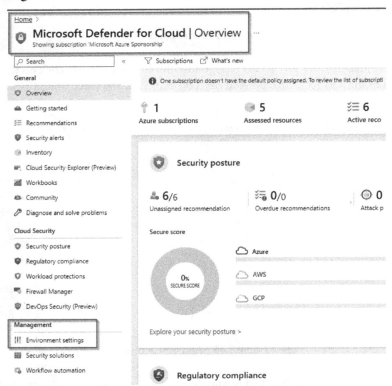

Figure 3.51 – Defender for Cloud

4. Click on the subscription you enabled for Microsoft Defender for Cloud earlier in this recipe:

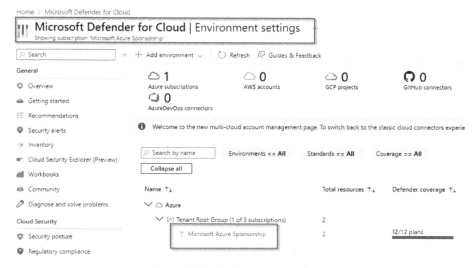

Figure 3.52 – Environment settings

5. From the **Defender plans** page, from the **Status** column, select **Off** for each **Plan** item:

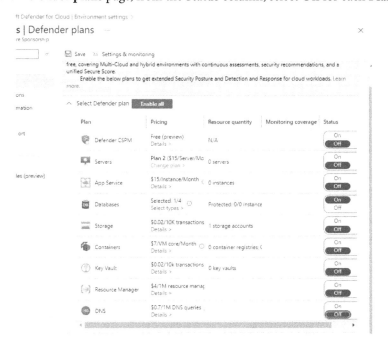

Figure 3.53 – Defender for Cloud plans

6. From the top menu bar, click **Save**, and click **Confirm** to downgrade:

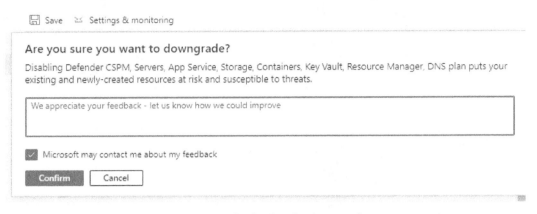

Figure 3.54 – Confirming the downgrade

7. You will receive a success notification:

Figure 3.55 – Success notification

This task of cleaning up the resources created in this recipe is complete.

How it works...

For this recipe, we looked at implementing JIT VM access to limit access to a VM's vulnerable management ports.

We set up an Azure VM resource that needed protection for remote access. We then enabled JIT access on our VM, looked at how to request access to a JIT access-enabled VM, and looked at auditing the activities.

There's more...

In this recipe, we covered enabling JIT VM access from within the **Virtual machines** blade in the Azure portal. We will look at enabling JIT VM access using **Defender for Cloud** (before November 2021, this was known as *Azure Security Center*) in *Chapter 8, Using Microsoft Defender for Cloud.*

The permissions needed to use and configure JIT access are covered in the following Microsoft Learn article:

- `https://learn.microsoft.com/en-us/azure/defender-for-cloud/just-in-time-access-overview?tabs=defender-for-container-arch-aks#what-permissions-are-needed-to-configure-and-use-jit`

See also

Should you require further information, you can refer to the following Microsoft Learn articles:

- `https://learn.microsoft.com/en-us/azure/defender-for-cloud/just-in-time-access`
- `https://learn.microsoft.com/en-us/azure/defender-for-cloud/defender-for-servers-introduction#defender-for-servers-plans`

4
Securing Virtual Machines

In the previous chapter, we covered recipes that allow you to provide secure remote access to Azure resources and minimize exposure to management ports on Azure **Virtual Machines** (**VMs**).

When we create Azure VMs or any Azure resource, we should take a **defense-in-depth** (**DiD**) approach. This means we should not rely on just the identity or network and remote access layers to secure our resources. We should, in addition, also apply protection controls at the *resource layer*, often referred to as *workload protection*.

This chapter will teach you how to secure and protect Azure VMs. We will break down the chapter into sections, covering using the *VM Update Management service* and protection through the *Microsoft Antimalware service* and *disk encryption*.

By the end of this chapter, you will have gained skills for securing Azure VMs through the following recipes:

- Implementing VM Update Management
- Implementing VM Microsoft Antimalware
- Implementing VM Azure Disk Encryption

Technical requirements

For this chapter, it is already assumed that you have an *Azure AD tenancy* and an *Azure subscription* from completing the recipes in previous chapters of this cookbook. If you skipped straight to this chapter, the information to create a new *Azure AD tenancy* and an *Azure subscription* for these recipes is included in the following list of requirements.

For this chapter, the following is required for the recipes:

- A device with a browser, such as Edge or Chrome, to access the Azure portal (`https://portal.azure.com`)

- An **Azure AD tenancy** and **Azure subscription**; you may use an existing subscription or sign up for free: `https://azure.microsoft.com/en-us/free`

- An **Owner role** for the **Azure subscription**

Implementing VM Update Management

Prevention is always better than cure. And so, it is always important to ensure the continued integrity of the software running on our VMs and to minimize the risk of a vulnerability being exploited.

Azure's VM **Update Management** is part of the **Azure Automation** solution and can aid in managing the complex operations of tracking and remediating software patching for *Azure* and *non-Azure* VMs.

This recipe will teach you how to implement VM **Update Management** as part of **Azure Automation** for your Azure VMs.

Getting ready

This recipe requires the following:

- A device with a browser, such as Edge or Chrome, to access the Azure portal (`https://portal.azure.com`).

- Access to an Azure subscription, where you have access to the **Owner** role for the **Azure subscription**.

- An Azure **Automation account** to manage VMs; we will step through creating an Automation account as a *Getting ready* task.

- A *Windows Server* **Azure VM** to use with this recipe; we will step through creating this VM as a *Getting ready* task.

Continue with the following *Getting ready* tasks for this recipe:

- Creating an Automation account

- Creating a VM

Getting ready task – creating an Automation account

Perform the following steps:

1. Sign in to the *Azure portal*: `https://portal.azure.com`.

2. From the top menu of the Azure portal, click **Create a resource**:

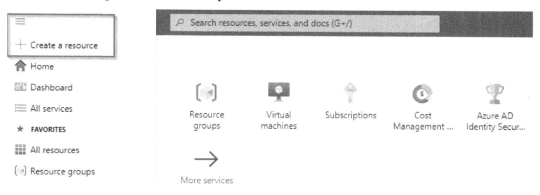

Figure 4.1 – Create a resource

3. On the **Create a resource** screen, in the search box, type automation, select the **Automation** tile from the results, click **Create**, and click **Automation**:

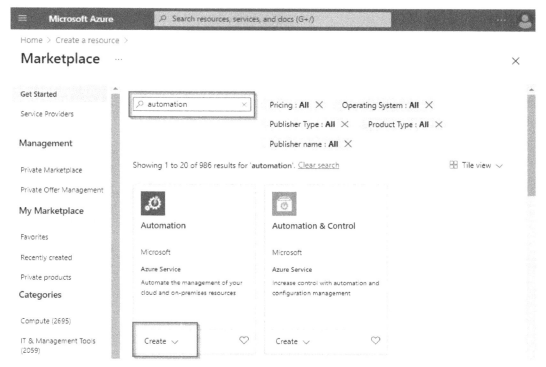

Figure 4.2 – Creating an Automation account

4. From the **Basics** tab of the **Create an Automation Account** page, under the **Project details** section, set the **Subscription** type as required; and for the **Resource group** type, select **Create new**, enter a **Name** value, and click **OK**.

5. Under the **Instance details** section, set as required the **Automation account name** and **Region** values.

6. Click **Review + create**.

7. Click **Create** on the **Review + create** tab.

8. A notification will display that the resource deployment succeeded.

This *Getting ready* task is complete. Next, we will create a VM for use with this recipe.

Getting ready task – creating a VM

Perform the following steps:

1. In the search box in the *Azure portal*, type `virtual machines`, and select **Virtual machines** from the listed **Services** results.

2. Click **+ Create** from the *top-left menu bar* on the **Virtual machine** screen and select **Azure virtual machine**.

3. From the **Basics** tab, under the **Project details** section, set the **Subscription** type as required. For the **Resource group** type, select the same *resource group* we created for the *Automation account*.

4. Under **Instance details**, set the following:

    *   **Virtual machine name**: *Type a name.*
    *   **Region**: *Select a region.* VMs can exist in any region, irrespective of the Automation account region.
    *   **Availability options**: Select **No infrastructure redundancy required**.
    *   **Security type**: Select **Standard**.
    *   **Image**: Select **Windows Server 2019 Datacenter – Gen2**.
    *   **Size**: *Leave the default (or set it as required to reduce recipe costs).*

5. Under **Administrator account**, set **Username** and **Password** values as required.

6. Click **Review + create**.

7. Click **Create** on the **Review + create** tab once *validation* has passed.

8. A notification will display that the resource deployment succeeded.

9. Click on **Go to resource** to open the VM page ready for the first task in this recipe.

The *Getting ready* tasks for this recipe are complete.

You are now ready to continue the main task for this recipe of enabling VM Update Management.

How to do it...

This recipe consists of the following task:

* Enabling Update Management from a VM

Task – enabling Update Management from a VM

Perform the following steps:

1. If you have not done so, navigate to the *VM* we created in the *Getting ready* task.

2. From the **VM** page, click **Updates** under the **Operations** section:

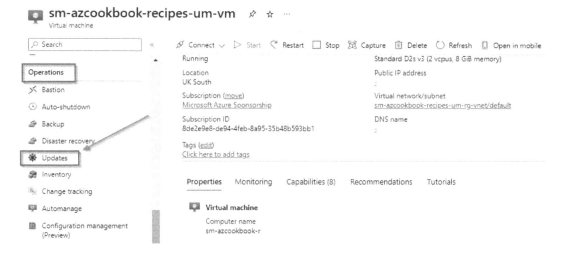

Figure 4.3 – Created VM page

3.   From the **Updates** page, under the **Updates using Automation** section, click **Go to Updates using automation**:

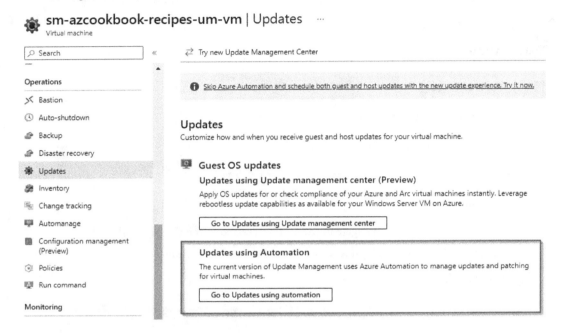

Figure 4.4 – VM Updates page

4.   From the **Update Management** page, leave **Log Analytics** settings as the *default*, then select your **Automation account subscription** type as required. For the **Automation account** setting, please select the *Automation account* we created in the *Getting ready* task, then click **Enable**:

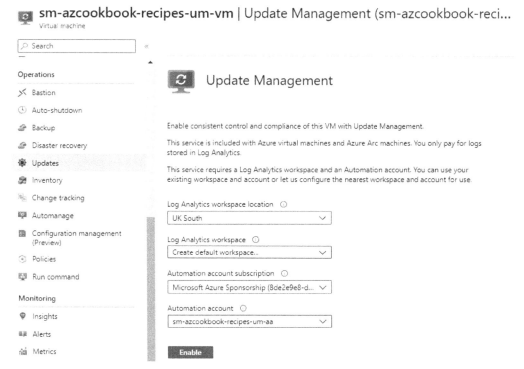

Figure 4.5 – Update Management settings

5.   The *Update Management solution* will now begin the deployment:

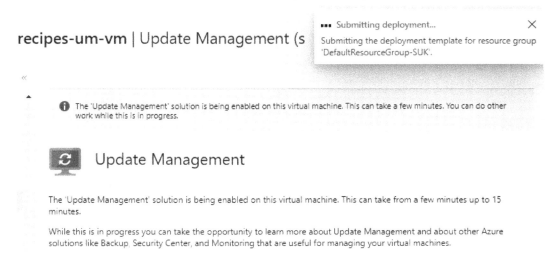

Figure 4.6 – Update Management deployment

6. You will receive a notification that the deployment succeeded:

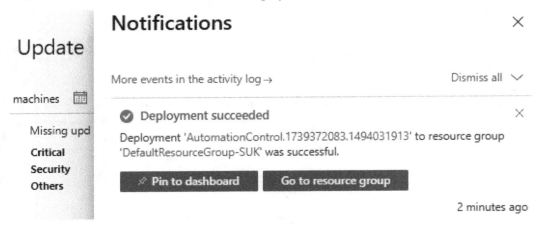

Figure 4.7 – Deployment succeeded

7. You can now start operating the *Update Management* solution:

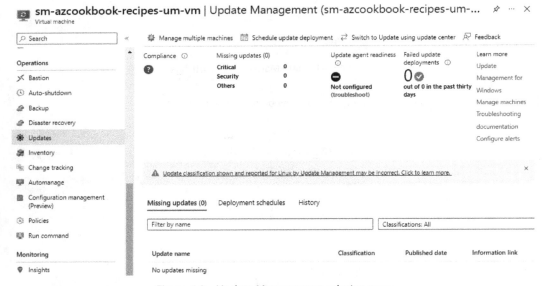

Figure 4.8 – Update Management solution page

This task to enable Update Management from a VM is now completed. In the next task, we'll clean up the resources created in this recipe.

Task – cleaning up resources

Perform the following steps:

1. In the search box in the *Azure portal*, type `resource groups` and select **Resource Groups** from the listed **Services** results.

2. From the **Resource groups** page, select the *resource group* we created for this recipe, and click **Delete resource group**; this will delete all the resources created as part of this recipe:

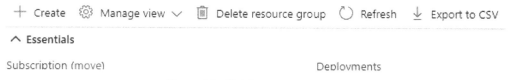

Figure 4.9 – Delete resource group

This task, to clean up the resources created in this recipe, is complete.

How it works...

For this recipe, we looked at enabling the Update Management service, which provides a fully automated and monitored software update solution for Azure VMs. We looked at the scenario for enabling Update Management directly from an Azure VM.

Alternatively, if you need to enable Update Management at scale across many VMs, you can use an alternative method of enabling it from an *Automation account*. This method is described in the following *Microsoft Learn* article: `https://learn.microsoft.com/en-us/azure/automation/update-management/enable-from-automation-account`.

There's more...

To operate the implemented Update Management solution, you can refer to the following *Microsoft Learn* articles:

- *Manage updates and patches for your VMs*: `https://learn.microsoft.com/en-us/azure/automation/update-management/manage-updates-for-vm`

- *Troubleshoot Update Management issues*: `https://learn.microsoft.com/en-us/azure/automation/troubleshoot/update-management`

In our recipe, all required permissions were assigned and inherited from the *Owner* role for our subscription. Working on the least privileges principle, you should only assign users the least access to perform their tasks in real-world environments.

Please refer to the following *Microsoft Learn* articles for more information:

- *Permissions for enabling Update Management, and Change Tracking and Inventory from a VM*: `https://learn.microsoft.com/en-us/azure/automation/automation-role-based-access-control#feature-setup-permissions`

- *Azure Automation account authentication overview*: `https://learn.microsoft.com/en-us/azure/automation/automation-security-overview`

See also

Should you require further information, you can refer to the following *Microsoft Learn* articles:

- *Security best practices for IaaS workloads in Azure*: `https://learn.microsoft.com/en-us/azure/security/fundamentals/iaas`

- *Azure Virtual Machines security overview*: `https://learn.microsoft.com/en-us/azure/security/fundamentals/virtual-machines-overview`

- *Azure Automation documentation*: `https://learn.microsoft.com/en-us/azure/automation`

- *Manage updates and patches for your VMs*: `https://learn.microsoft.com/en-us/azure/automation/update-management/manage-updates-for-vm`

- *Configure Update Management*: `https://learn.microsoft.com/en-us/training/modules/host-security/7-update-management`

Implementing VM Microsoft Antimalware

In implementing our *DiD* strategy, we have looked at recipes to protect our identities and networks and remediate any unpatched Azure VMs on the network.

Our next level of defense is a need for a security capability that will protect our VMs in real time from **malicious software** (**malware**) such as **viruses**, **worms**, and **Trojans**.

Microsoft Antimalware is a free built-in solution that offers a protection capability that alerts and remediates these malware threats in real time.

Getting ready

This recipe requires the following:

- A device with a browser, such as Edge or Chrome, to access the Azure portal (`https://portal.azure.com`)

- Access to an Azure subscription, where you have access to the **Owner** role for the **Azure subscription**

How to do it...

This recipe consists of the following task:

- Enabling Antimalware when creating a VM

Task – enabling Antimalware when creating a VM

Perform the following steps:

1. Sign in to the *Azure portal*: `https://portal.azure.com`.

2. In the search box in the *Azure portal*, type `virtual machines` and select **Virtual machines** from the listed **Services** results.

3. Click + **Create** from the *top-left menu bar* on the **Virtual machine** screen and select **Azure virtual machine**.

4. From the **Basics** tab, under the **Project details** section, set the **Subscription** type as required.

5. For the **Resource group** type, click **Create new**.

6. Enter a **Name** value and click **OK**.

7. Under **Instance details**, set the following:

   - **Virtual machine name**: *Type a name*

   - **Region**: *Select a region*

   - **Availability options**: Select **No infrastructure redundancy required**

   - **Security type**: Select **Standard**

   - **Image**: Select **Windows Server 2019 Datacenter – Gen2**

   - **Size**: *Leave the default (or set it as required to reduce recipe costs)*

8. Under **Administrator account**, set **Username** and **Password** values as required.

9. Click the **Advanced** tab to skip through the wizard to the configuration step we need for this recipe:

Create a virtual machine ...

Basics   Disks   Networking   Management   Monitoring   Advanced   Tags   Review + create

Add additional configuration, agents, scripts or applications via virtual machine extensions or cloud-init.

Extensions

Extensions provide post-deployment configuration and automation.

Extensions ⓘ   Select an extension to install

Figure 4.10 – Advanced tab

10. From the **Advanced** tab, from the **Extensions** section, click **Select an extension to install**:

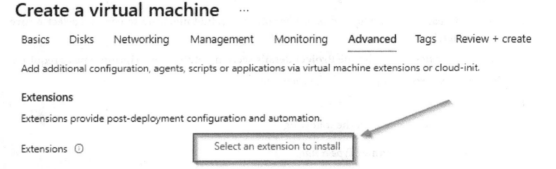

Figure 4.11 – Selecting extensions to install

11. From the **Install an Extension** page, type microsoft antimalware in the search box, select **Microsoft Antimalware** from the results, and then click **Next**:

Install an Extension   ...

microsoft antimalware   ✕

KeyVault for Windows

Microsoft Corp.

KeyVault Virtual Machine Extension

Microsoft Antimalware

Microsoft Corp.

Microsoft Antimalware for Azure Virtual Machines

Network Watcher Agent for Windows

Microsoft Corp.

Azure Network Watcher is a network performance monitoring, diagnostic and analytics service that enables you to monitor your network in Azure

NVIDIA GPU Driver Extension

Next

Microsoft Antimalware

Publisher: Microsoft Corp.

Overview

Microsoft Antimalware for Azure Virtual Machines is a real-time protection capability that helps identify and remove viruses, spyware, and other malicious software, with configurable alerts when known malicious or unwanted software attempts to install itself or run on your system. The solution can be enabled and configured from the Azure Portal, Service Management REST API, and Microsoft Azure PowerShell SDK cmdlets.

To **enable** antimalware with the **default configuration**, click **Create** on the Add Extension blade without inputting any configuration setting values.

To **enable** antimalware with a **custom configuration**, input the supported values for the configuration settings provided on the **Add Extension** blade and click **Create**. Please refer to the **tooltips** provided with each configuration setting on the Add Extension blade to see the supported configuration values.

To **enable antimalware event collection** for a virtual machine, click any part of the **Monitoring lens** in the virtual machine blade, click **Diagnostics** command on Metric blade, select **Status ON** and check **Windows Event system logs.** The antimalware events are collected from the Windows Event system logs to your storage account. You can configure the storage account for your virtual machine to collect the antimalware events by selecting the appropriate storage account.

Legal Terms

By clicking the Create button, I acknowledge that I am getting this software from Microsoft Corp. and that the legal terms of Microsoft Corp. apply to it. Microsoft does not provide rights for third-party software. Also see the privacy statement from Microsoft Corp..

Figure 4.12 – Selecting an extension

12. From the **Configure Microsoft Antimalware Extension** page, make the appropriate selections or leave the *defaults*, then click **Create**:

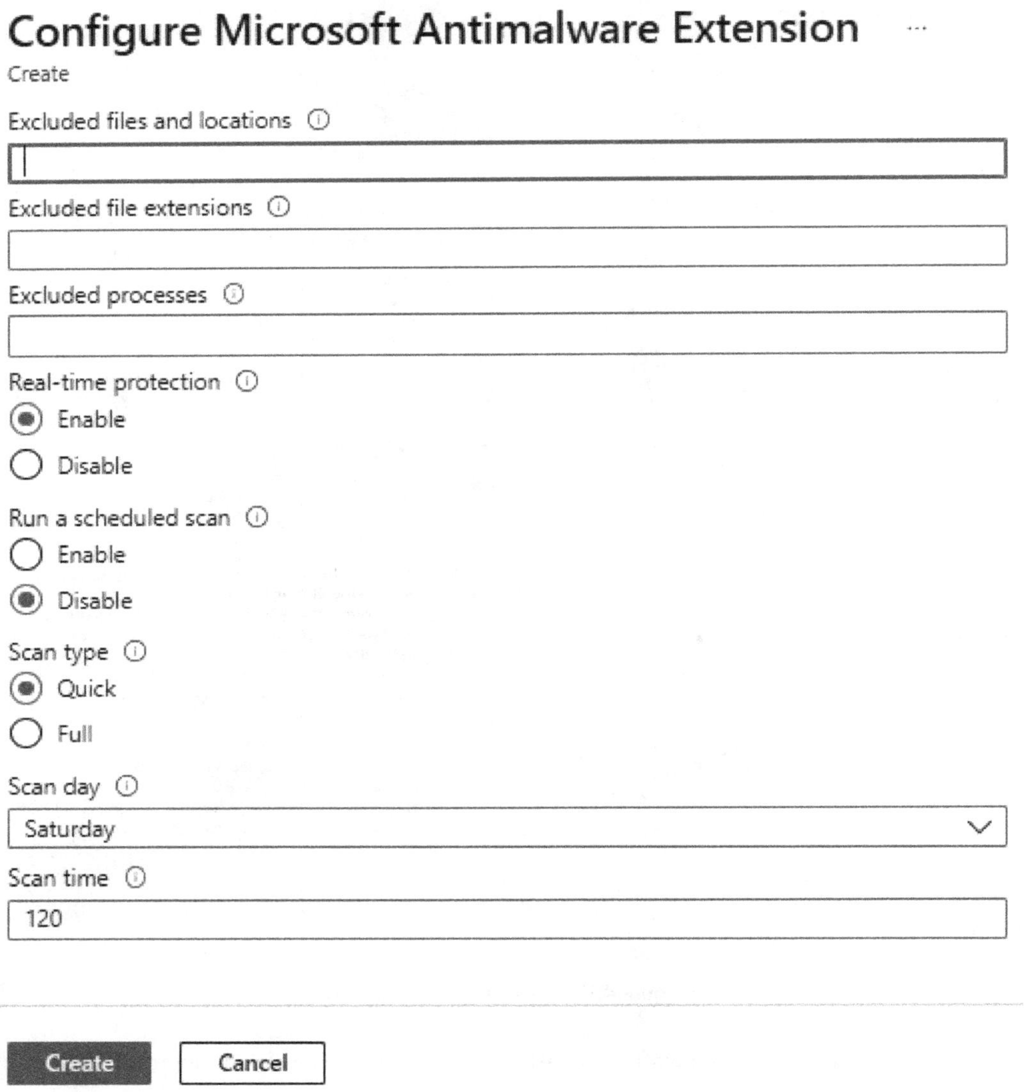

Configure Microsoft Antimalware Extension ···

Create

Excluded files and locations ⓘ

Excluded file extensions ⓘ

Excluded processes ⓘ

Real-time protection ⓘ

◉ Enable

○ Disable

Run a scheduled scan ⓘ

○ Enable

◉ Disable

Scan type ⓘ

◉ Quick

○ Full

Scan day ⓘ

Saturday ⌄

Scan time ⓘ

120

Create   Cancel

Figure 4.13 – Configuring the extension

13. Click **Review + create**.

14. Click **Create** on the **Review + create** tab once *validation* has passed.

15. A notification will display that the resource deployment succeeded.

This task to install Antimalware when creating a VM is now completed. In the next task, we'll clean up the resources created in this recipe.

Task – cleaning up resources

Perform the following steps:

1. In the search box in the *Azure portal*, type `resource groups` and select **Resource Groups** from the listed **Services** results.

2. From the **Resource groups** page, select the *resource group* we created for this recipe, and click **Delete resource group**; this will delete all the resources created as part of this recipe:

Figure 4.14 – Delete resource group

This task, to clean up the resources created in this recipe, is complete.

How it works...

For this recipe, we looked at implementing Microsoft Antimalware, which offers real-time protection against malware threats. The solution runs without human intervention as an automated and monitored background service.

We deployed the protection capabilities with the secure-by-default basic configuration; you may also implement more advanced custom configurations and Microsoft Defender for Endpoint for an even greater depth of protection.

See also

Should you require further information, you can refer to the following *Microsoft Learn* articles:

*   *Security best practices for IaaS workloads in Azure*: https://learn.microsoft.com/en-us/azure/security/fundamentals/iaas

*   *Azure Virtual Machines security overview*: https://learn.microsoft.com/en-us/azure/security/fundamentals/virtual-machines-overview

*   *Understand the malware threat*: https://learn.microsoft.com/en-us/training/modules/secure-vms-with-azure-security-center/4-malware-protection

*   *Microsoft Antimalware for Azure Cloud Services and Virtual Machines*: https://learn.microsoft.com/en-us/azure/security/fundamentals/antimalware

Implementing VM Azure Disk Encryption

Azure Disk Encryption (ADE) provides encryption of data on VM disks at **rest** in **Azure Storage**. The solution uses an integrated **Azure Key Vault** to store and manage the *encryption keys*.

Getting ready

This recipe requires the following:

- A device with a browser, such as Edge or Chrome, to access the Azure portal (`https://portal.azure.com`)
- You should sign in to an Azure subscription with the *Owner* role

Continue with the following *Getting ready* task for this recipe:

- Creating a VM

Getting ready task – creating a VM

Perform the following steps:

1. Sign in to the Azure portal: `https://portal.azure.com`.
2. In the search box in the *Azure portal*, type `virtual machines` and select **Virtual machines** from the listed **Services** results.
3. Click **+ Create** from the *top-left menu bar* on the **Virtual machine** screen and select **Azure virtual machine**.
4. From the **Basics** tab, under the **Project details** section, set the **Subscription** type as required.
5. For the **Resource group** type, click **Create new**.
6. Enter a **Name** value and click **OK**.
7. Under **Instance details**, set the following:

    - **Virtual machine name**: *Type a name*
    - **Region**: *Select a region*
    - **Availability options**: Select **No infrastructure redundancy required**
    - **Security type**: Select **Standard**
    - **Image**: Select **Windows Server 2019 Datacenter – Gen2**
    - **Size**: *Leave the default (or set it as required to reduce recipe costs)*

8. Under **Administrator account**, set **Username** and **Password** values as required.

9. Click **Next: Disks**, then **Next : Networking**, then **Next : Management**, and finally, **Next: Monitoring**.

10. From the **Monitoring** tab, check the **Enable OS guest diagnostics** box, and accept the *default name* for the new **Diagnostics storage account** entry:

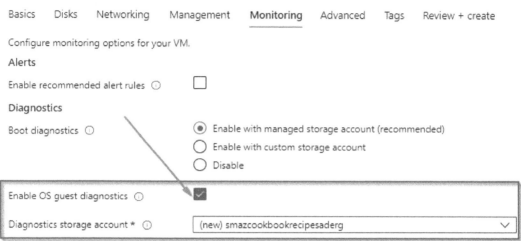

Figure 4.15 – Enabling guest diagnostics

11. Click **Review + create**.

12. Click **Create** on the **Review + create** tab once *validation* has passed.

13. A notification will display that the resource deployment succeeded.

14. Click on the **Go to resource** button once you are ready to start the task of *encrypting a VM*:

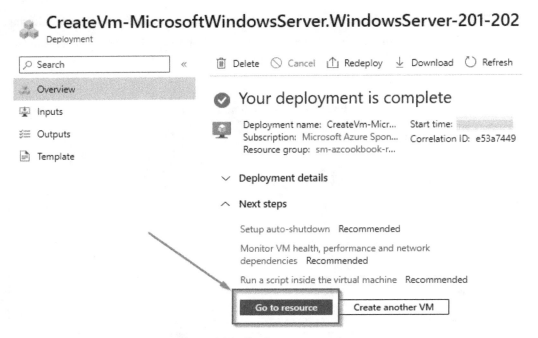

Figure 4.16 – Deployment complete

This *Getting ready* task, to create a VM for this recipe, is complete.

You are now ready to continue the main tasks for this recipe of encrypting a Windows VM using ADE.

How to do it...

This recipe consists of the following tasks:

- Encrypting a VM

Task – encrypting a VM

Perform the following steps:

1. From the screen of your VM in the *Azure portal*, click on **Disks** on the *left-hand sidebar* under the **Settings** section:

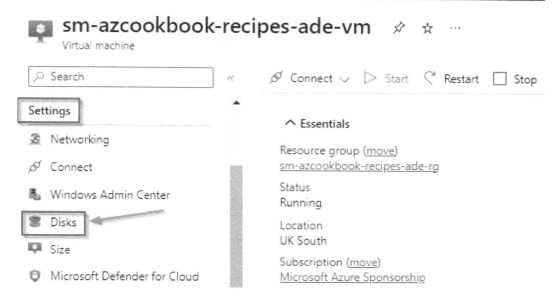

Figure 4.17 – VM page: disk selection

2. From the *top bar*, click **Additional settings**:

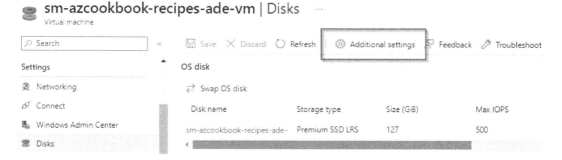

Figure 4.18 – Disks | Additional settings

3. From the **Disks** settings page, under the **Encryption settings** section, click the dropdown for **Disks to encrypt** and select **OS and data disks**:

Encryption settings

Azure Disk Encryption (ADE) provides volume encryption for the OS and data disks. Learn more about Azure Disk Encryption.

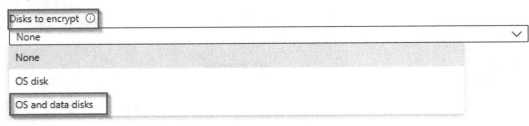

Figure 4.19 – Encryption settings

4.   From the **Key Vault** setting, click **Create a key vault**:

Encryption settings

Azure Disk Encryption (ADE) provides volume encryption for the OS and data disks. Learn more about Azure Disk Encryption.

Disks to encrypt ⓘ

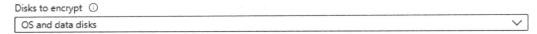

Azure Disk Encryption is integrated with Azure Key Vault to help manage encryption keys. As a prerequisite, you need to have an existing key vault with encryption permissions set. For additional security, you can create or choose an optional key encryption key to protect the secret.

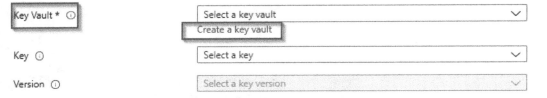

Figure 4.20 – Create a key vault

5.   From the **Create a key vault** page, on the **Basics** tab, under **Instance details**, enter as required a **Key vault name** value; this must be a *unique name*:

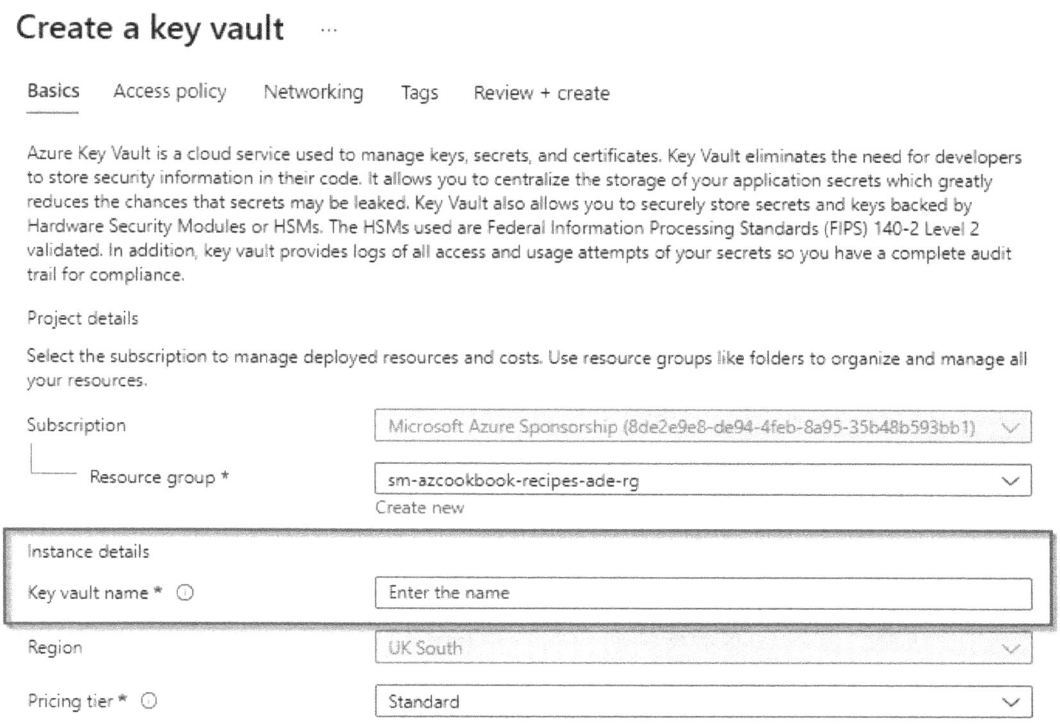

Create a key vault ...

Basics   Access policy   Networking   Tags   Review + create

Azure Key Vault is a cloud service used to manage keys, secrets, and certificates. Key Vault eliminates the need for developers to store security information in their code. It allows you to centralize the storage of your application secrets which greatly reduces the chances that secrets may be leaked. Key Vault also allows you to securely store secrets and keys backed by Hardware Security Modules or HSMs. The HSMs used are Federal Information Processing Standards (FIPS) 140-2 Level 2 validated. In addition, key vault provides logs of all access and usage attempts of your secrets so you have a complete audit trail for compliance.

Project details

Select the subscription to manage deployed resources and costs. Use resource groups like folders to organize and manage all your resources.

Subscription	Microsoft Azure Sponsorship (8de2e9e8-de94-4feb-8a95-35b48b593bb1) ⌄
Resource group *	sm-azcookbook-recipes-ade-rg ⌄
	Create new

Instance details

Key vault name * ⓘ	Enter the name

Region	UK South ⌄
Pricing tier * ⓘ	Standard ⌄

Figure 4.21 – Key vault settings

6. Click **Next** to move to the **Access policy** tab.

7. From the **Access policy** tab, under the **Resource access** section, check the **Azure Disk Encryption for volume encryption** box:

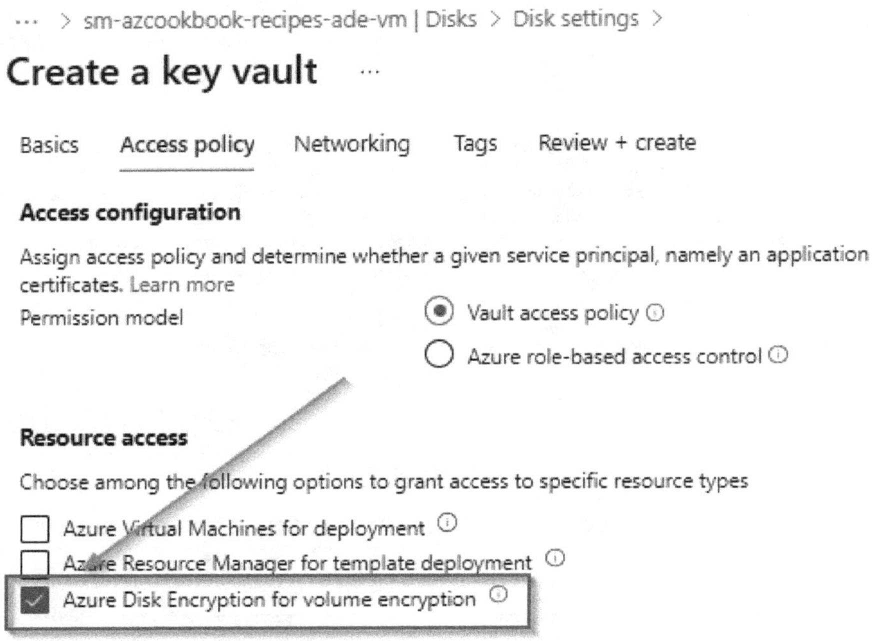

··· > sm-azcookbook-recipes-ade-vm | Disks > Disk settings >

Create a key vault   ···

Basics   **Access policy**   Networking   Tags   Review + create

Access configuration

Assign access policy and determine whether a given service principal, namely an application certificates. Learn more

Permission model   ◉ Vault access policy ⓘ

                                     ○ Azure role-based access control ⓘ

Resource access

Choose among the following options to grant access to specific resource types

☐ Azure Virtual Machines for deployment ⓘ

☐ Azure Resource Manager for template deployment ⓘ

☑ Azure Disk Encryption for volume encryption ⓘ

Figure 4.22 – Access policy

8. Click **Review + create**.

9. Click **Create** on the **Review + create** tab.

10. Once the *key vault* has been successfully created, you will be returned to the **Disk settings** page.

11. For the **Key** setting, click **Create a key**:

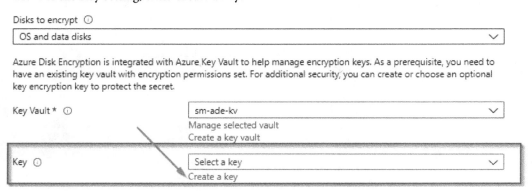

Figure 4.23 – Create a key

12. From the **Create a key** page, enter the required **Name** value, then click **Create**:

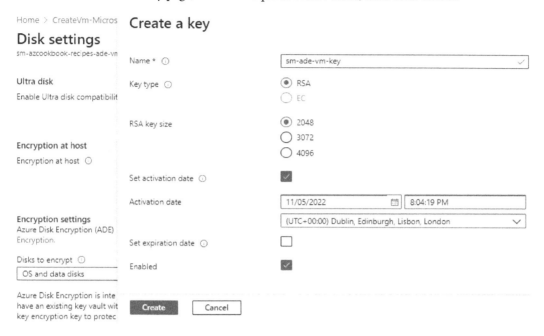

Figure 4.24 – Key settings

13. Once the *key* has been successfully created, you will be returned to the **Disk settings** page.

14. Click **Save**.

15. A notification will display that the disk encryption settings succeeded:

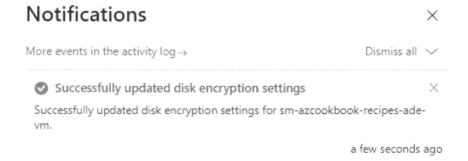

Figure 4.25 – Settings succeeded

16. The encryption status of the VM disk volumes can be verified by running the following **PowerShell** command from Azure Cloud Shell:

```
Get-AzVmDiskEncryptionStatus -VMName MyVM
-ResourceGroupName MyResourceGroup
```

The following screenshot shows the output of this command:

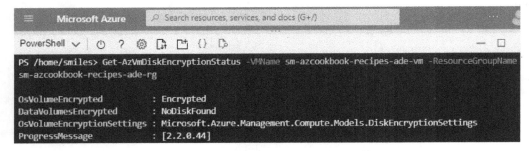

Figure 4.26 – Encryption status

This task of encrypting a Windows VM using ADE is complete. In the next task, we'll clean up the resources created in this recipe.

Task – cleaning up resources

Perform the following steps:

1. From Azure Cloud Shell, run the following PowerShell command:

```
Remove-AzResourceGroup -Name "myResourceGroup"
```

2. Type Y or click *Enter* to accept removing the *resource group*:

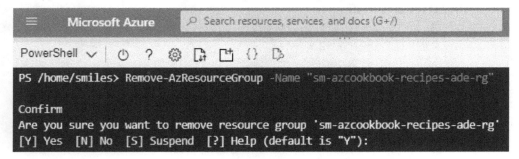

Figure 4.27 – Deleting a resource group

This task to clean up the resources created in this recipe is complete.

How it works...

For this recipe, we looked at implementing ADE using the Azure portal to provide encryption of data on VM disks at rest in Azure Storage. We created a VM and a key vault configured to store encryption keys and then encrypted the OS disk of the VM. We verified the OS disk had been encrypted with ADE using the following PowerShell command run from Cloud Shell:

```
Get-AzVmDiskEncryptionStatus
```

ADE provides *volume-level encryption* for all OSs and data disks attached to a VM and is *zone resilient*. ADE uses the *Windows BitLocker* feature, but you do not directly interact with BitLocker or use BitLocker to decrypt a VM that used ADE to perform the encryption.

The supported VMs and OSs are listed as follows:

- Generation 1 and 2 VMs

- VMs with SSD premium storage are also supported

- VMs with more than 2 GB of memory

- Client OS: Windows 8 and newer, including Windows 10 Enterprise multi-session

- Server OS: Windows Server 2008 R2 and newer

Please refer to the following *Microsoft Learn* article for unsupported scenarios:

```
https://learn.microsoft.com/en-us/azure/virtual-machines/windows/
disk-encryption-windows#unsupported-scenarios
```

You must ensure the following for the Windows VM to have ADE configured:

- Must be able to connect to the `login.microsoftonline.com` Azure AD endpoint

- Must be able to connect to the Azure Key Vault endpoint; please refer to `https://learn.microsoft.com/en-us/azure/key-vault/general/access-behind-firewall`

- Must be able to connect to an Azure Storage endpoint that hosts the Azure VM

There's more...

In addition to ADE as a type of encryption available for managed disks, there are other options available, as follows:

- Azure **Storage Service Encryption (SSE)**

- Encryption at the host

- Confidential disk encryption

The core difference between ADE and SSE is that ADE operates at the VM **virtual hard disk (VHD)** level, and only the VM that owns the disk can access the encrypted disk image. SSE operates at the physical disk level when the data is decrypted and loaded into memory when the data on the disk is accessed.

Further information on these additional encryption capabilities can be found in the following *Microsoft Learn* articles:

- *Security best practices for IaaS workloads in Azure*: https://learn.microsoft.com/en-us/azure/security/fundamentals/iaas

- *Azure Virtual Machines security overview*: https://learn.microsoft.com/en-us/azure/security/fundamentals/virtual-machines-overview

- *Overview of managed disk encryption options*: https://learn.microsoft.com/en-us/azure/virtual-machines/disk-encryption-overview

- Encryption options for protecting Windows and Linux VMs: https://learn.microsoft.com/en-us/training/modules/secure-your-azure-virtual-machine-disks/2-encryption-options-for-protecting-windows-and-linux-vms

See also

Should you require further information, you can refer to the following *Microsoft Learn* articles:

- *Azure Disk Encryption for Windows VMs*: https://learn.microsoft.com/en-us/azure/virtual-machines/windows/disk-encryption-overview

- *Secure your Azure virtual machine disks*: https://learn.microsoft.com/en-us/training/modules/secure-your-azure-virtual-machine-disks

5
Securing Azure SQL Databases

In the previous chapter, we covered recipes for protecting the integrity of Azure VMs by ensuring that they were *updated*, *antimalware* was enabled, and disks were *encrypted* disks.

With public cloud provider platforms, the shared responsibility model means that while the provider is responsible for providing security and control mechanisms of the platform hosting that data that can be enabled, the customer is **always** responsible for correctly implementing and configuring those controls and ensuring appropriate governance and operations.

To avoid doubt, it is critical to call out that the customer is **always** responsible for the data stored on those platforms and its operation.

We can use an analogy of a rented property and your relationship with the landlord. While the landlord will be responsible for providing doors and windows and the controls such as locks, alarms, and a gated entrance for vehicle access with CCTV to monitor the property, you are responsible for ensuring that you have shut the doors, set the alarm, closed the gates, and operated the CCTV.

If you are not enabling and configuring these controls from the platform provider to secure your data, you are negligent in your duty of care for that data.

This chapter will teach you how to secure and protect Azure databases.

By the end of this chapter, you will have covered the following recipes to secure Azure databases:

- Implementing a service-level IP firewall
- Implementing a private endpoint
- Implementing Azure AD authentication and authorization

If you wish to have a primer on securing databases or learn additional functionality, you can read the following Microsoft articles:

- What is database security?: `https://azure.microsoft.com/en-us/resources/cloud-computing-dictionary/what-is-database-security/#what-is-database-security`
- Configure and manage SQL database security: `https://learn.microsoft.com/en-us/training/modules/sql-database-security`
- Auditing for Azure SQL Database and Azure Synapse Analytics: `https://learn.microsoft.com/en-us/azure/azure-sql/database/auditing-overview`
- Always Encrypted documentation: `https://learn.microsoft.com/en-us/azure/azure-sql/database/always-encrypted-landing`

Technical requirements

For this chapter, it is already assumed that you have an *Azure AD tenancy* and an *Azure subscription* from completing the recipes in previous chapters of this cookbook. If you skipped straight to this section, the information to create a new *Azure AD tenancy* and an *Azure subscription* for these recipes is included in the following list of requirements.

For this chapter, the following are required for the recipes:

- A device with a browser, such as Edge or Chrome, to access the Azure portal: `https://portal.azure.com`
- An **Azure AD tenancy** and **Azure subscription**; you may use an existing one or sign up for free: `https://azure.microsoft.com/en-us/free`
- An **Owner role** for the **Azure subscription**

Implementing a service-level IP firewall

In many workload scenarios, the first line of protection in a *defense-in-depth* approach to security is a **network layer firewall** to act as a *layer 3* network traffic *packet filter*.

This recipe will teach you how to restrict network access to your Azure SQL database. We will configure rules for the native Azure *service-level* IP firewall service to protect your Azure databases.

Getting ready

This recipe requires the following:

- A device with a browser, such as Edge or Chrome, to access the Azure portal: `https://portal.azure.com`

- Access to an Azure subscription, where you have access to the **Owner** role for the **Azure subscription**

- Access to an **Azure SQL database**; we will step through this process in the following *Getting ready* tasks

Continue with the following *Getting ready* tasks for this recipe:

- Creating an Azure SQL database

Getting ready task – creating an Azure SQL database

Perform the following steps:

1. Sign in to the Azure portal: `https://portal.azure.com`.

2. From the top menu of the Azure portal, in the **Search** box, type `SQL databases`, and click on **SQL databases** from the results:

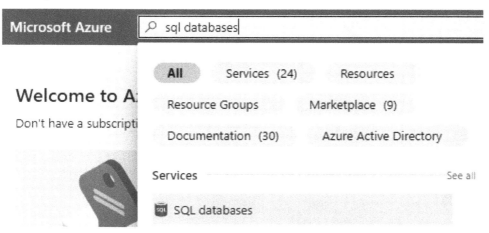

Figure 5.1 – Searching for a resource

3. Click + **Create** from the *top toolbar* from the **SQL databases** page.

4. From the **Basics** tab of the **Create SQL Database** page, under the **Project details** section, set your **Subscription** as required. Then, under **Resource group**, select **Create new**, enter a **Name**, and click **OK**.

5. Under the **Database details** section, enter a **Database name** as required, and for **Server**, click **Create new**.

6. From the **Create SQL Database Server** page, under the **Server details** section, set a **Server name** (*this must be globally unique*) and **Location** as required:

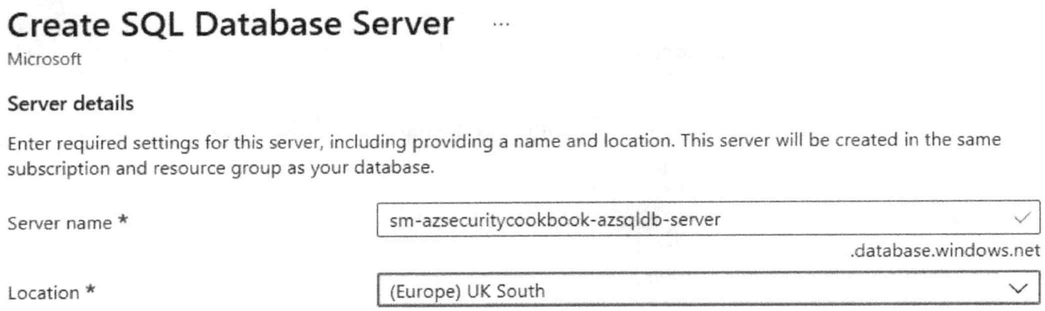

Create SQL Database Server ⋯
Microsoft

Server details

Enter required settings for this server, including providing a name and location. This server will be created in the same subscription and resource group as your database.

Server name *	sm-azsecuritycookbook-azsqldb-server ✓
	.database.windows.net
Location *	(Europe) UK South ∨

Figure 5.2 – Create SQL Database Server

7. Under the **Authentication** section, set **Authentication method** to **Use SQL authentication**, and then enter a **Server admin login** and **Password** as required. Then, click **OK**:

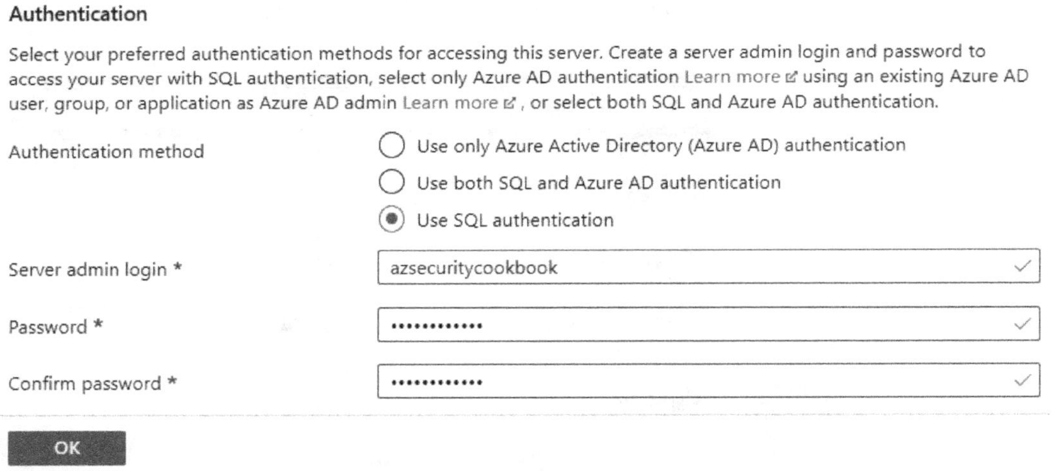

Authentication

Select your preferred authentication methods for accessing this server. Create a server admin login and password to access your server with SQL authentication, select only Azure AD authentication Learn more ☑ using an existing Azure AD user, group, or application as Azure AD admin Learn more ☑ , or select both SQL and Azure AD authentication.

Authentication method	◯ Use only Azure Active Directory (Azure AD) authentication
	◯ Use both SQL and Azure AD authentication
	⦿ Use SQL authentication
Server admin login *	azsecuritycookbook ✓
Password *	••••••••••• ✓
Confirm password *	••••••••••• ✓

> OK

Figure 5.3 – Setting up authentication

8. Set **Workload environment** to **Development** to reduce running costs for this recipe:

Workload environment ⦿ Development
◯ Production

Figure 5.4 – Setting the environment

9. From the **Networking** tab, under the **Network connectivity** section, set **Connectivity method** to **Public Endpoint**:

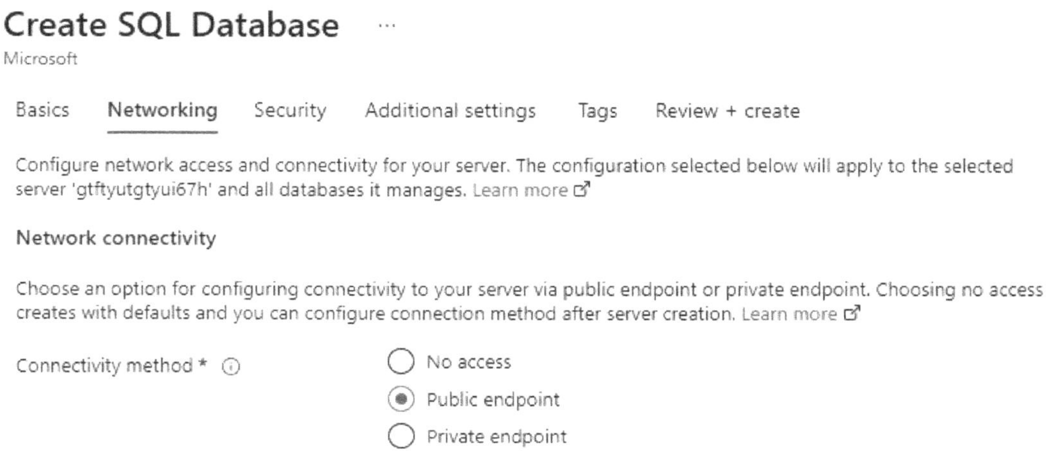

Create SQL Database   ···

Microsoft

Basics   **Networking**   Security   Additional settings   Tags   Review + create

Configure network access and connectivity for your server. The configuration selected below will apply to the selected server 'gtftyutgtyui67h' and all databases it manages. Learn more ⬀

Network connectivity

Choose an option for configuring connectivity to your server via public endpoint or private endpoint. Choosing no access creates with defaults and you can configure connection method after server creation. Learn more ⬀

Connectivity method * ⓘ

◯ No access
⦿ Public endpoint
◯ Private endpoint

Figure 5.5 – Setting network connectivity

10. For this recipe, no other configuration needs to be reviewed or required; click **Review + Create**.

11. Click **Create** on the **Review + create** tab.

12. A notification will display that the resource deployment succeeded.

13. Click on **Go to resource** so that you're ready for the first task for this recipe.

This *Getting ready* task is complete. You are now ready to continue the main tasks for this recipe, which involve setting a service-level IP firewall.

How to do it...

This recipe consists of the following tasks:

- Setting server-level firewall rules
- Setting database-level firewall rules
- Cleaning up resources

Task – setting server-level firewall rules

Perform the following steps:

1. From the created **Azure SQL database** page, click **Overview** from the top of the left toolbar:

Figure 5.6 – Configuring access

2. From the **Overview** page, click **Set server firewall** from the *top toolbar*:

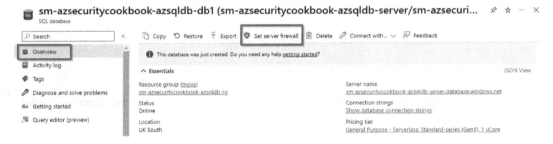

Figure 5.7 – Set server firewall

3. From the **Public access** tab, under **Public network access**, click **Selected networks**:

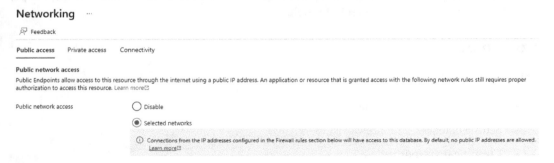

Figure 5.8 – Selected networks

4. Under **Firewall rules**, click + **Add a firewall rule**:

Firewall rules

Allow certain public internet IP addresses to access your resource. Learn more⬈

+ Add your client IPv4 address (90.152.127.206)   + Add a firewall rule

Rule name	Start IPv4 address	End IPv4 address	
ClientIPAddress_2022-11-18_10-52-50	90.152.127.206	90.152.127.206	🗑

Figure 5.9 – Add a firewall rule

5. Add **Rule name**, **Start IP**, and **End IP** for your scenario:

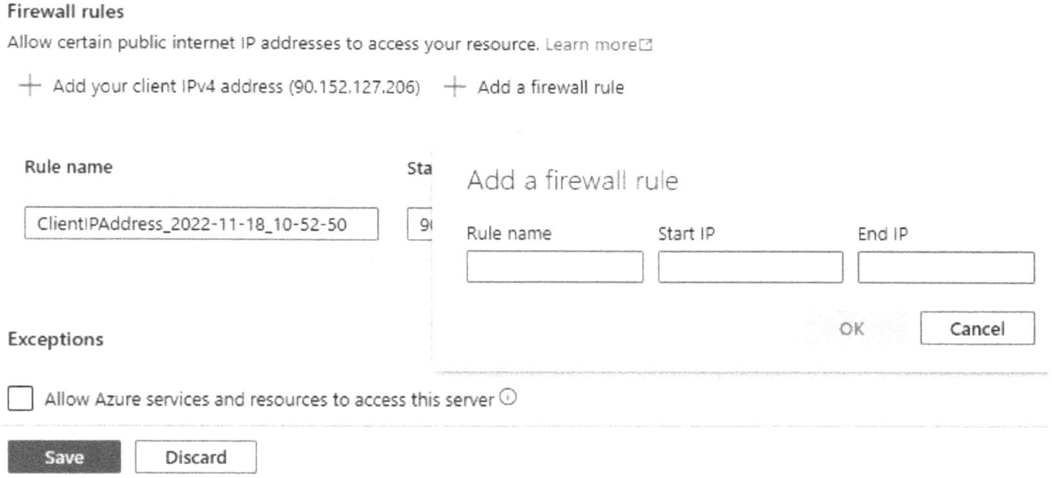

Firewall rules

Allow certain public internet IP addresses to access your resource. Learn more⬈

+ Add your client IPv4 address (90.152.127.206)   + Add a firewall rule

Rule name	Sta	Add a firewall rule		
ClientIPAddress_2022-11-18_10-52-50	9	Rule name	Start IP	End IP

Exceptions

☐ Allow Azure services and resources to access this server ⓘ

OK   Cancel

Save   Discard

Figure 5.10 – Setting the firewall information

6. Click **Save**.

With that, you have set a server-level firewall rule. In the next task, we will set a database-level firewall rule.

Task – setting database-level firewall rules

Perform the following steps:

1. From the created **Azure SQL database** page, click **Query editor** from the left menu:

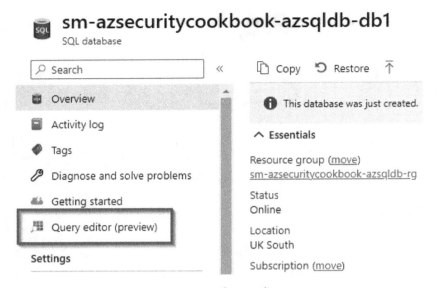

Figure 5.11 – Query editor

2. Enter your **SQL server authentication** credentials from the **Query Editor** page, then click **OK**:

Figure 5.12 – Authenticating to the database

3. From the **Query Editor** area, **enter** EXECUTE sp_set_database_firewall_rule
    N'Example Cookbook DB Rule', '<your-device_publicIP>','<your-
    device_publicIP>'; as a query and hit **Run**:

Figure 5.13 – Running a query

4. From the *top toolbar* of the **Query Editor** area, click on + **New Query**.

5. From the newly opened **Query** pane, **enter** SELECT * FROM sys.database_firewall_
    rules ORDER BY name; as a query and hit **Run**:

6. You will see the rule you created in the **Results** pane:

id	name	start_ip_address	end_ip_address
1	Example Cookbook DB Rule	90.152.127.206	90.152.127.206

Figure 5.15 – Viewing the query's result

With that, you have set a database-level firewall rule. In the next task, we will clean up the resources that were created in this recipe.

Task – cleaning up resources

Perform the following steps:

1. From the **Search box** area in the Azure portal, type `resource groups` and select **Resource Groups** from the listed **Services** results.

2. From the **Resource groups** page, select the *resource group* we created for this recipe and click **Delete resource group**; this will delete all the resources that were created as part of this recipe:

Figure 5.16 – Delete resource group

With that, you have cleaned up the resources that were created in this recipe.

How it works...

For this recipe, we looked at implementing a service-level IP firewall. This allowed us to remove access from all networks and then explicitly define the public internet IP addresses where we wish to access our SQL servers and databases for remote access purposes by administrators.

As a getting ready task, we created an Azure SQL database to illustrate the public PaaS service we want to restrict network access to.

We wanted to show you how to ensure that connections can only be made from those networks and specific IPs we explicitly allow.

See also

Should you require further information, you can refer to the following Microsoft Learn articles:

-   An overview of Azure SQL Database and SQL Managed Instance security capabilities: `https://learn.microsoft.com/en-us/azure/azure-sql/database/security-overview`

-   Outbound firewall rules for Azure SQL Database and Azure Synapse Analytics: `https://learn.microsoft.com/en-us/azure/azure-sql/database/outbound-firewall-rule-overview`

-   Azure SQL Database and Azure Synapse IP firewall rules: `https://learn.microsoft.com/en-us/azure/azure-sql/database/firewall-configure`

- Exercise – Restrict network access: `https://learn.microsoft.com/en-us/training/modules/secure-your-azure-sql-database/2-restrict-network-access`

- Azure SQL Database and Azure Synapse Analytics network access controls: `https://learn.microsoft.com/en-us/azure/azure-sql/database/network-access-controls-overview`

- Quickstart: Use SSMS to connect to and query an Azure SQL database or Azure SQL Managed Instance: `https://learn.microsoft.com/en-us/azure/azure-sql/database/connect-query-ssms`

Implementing a private endpoint

One of the foundations of securing resources is to reduce the attack surface area and minimize exposure to public network access.

Azure PaaS services' inherent nature and concern are that they have public endpoints, which means any vulnerabilities are exposed and can be exploited. One of the best practices we should follow in our security model for cloud services is to limit public access to resources wherever possible and adopt a network model where private IP addressing is used wherever possible.

Azure Private Link is an Azure service that enables you to connect to *public endpoint PaaS services* such as **Azure SQL database** and **Azure Storage** from an *Azure virtual network.*

Using the **Private Link** capability, you can privately connect to a PaaS service by creating a private endpoint (*in place of a public endpoint*) to connect to.

The PaaS service endpoint is never exposed to the internet, and traffic to and from the service never traverses the public internet. As such, traffic stays private on the Microsoft backbone network.

This recipe will teach you how to implement a private endpoint to provide non-public secure endpoint access to your Azure databases.

Getting ready

This recipe requires the following:

- A device with a browser, such as Edge or Chrome, to access the Azure portal: `https://portal.azure.com`

- Access to an Azure subscription, where you have access to the **Owner** role for the **Azure subscription**

- Access to an **Azure SQL database** instance; we will step through creating this as a *Getting ready* task

- An **Azure Virtual Network**; we will step through creating this as a *Getting ready* task

Continue with the following getting ready tasks for this recipe:

- Creating an Azure Virtual Network
- Creating an Azure SQL database

Getting ready task – creating an Azure Virtual Network

Perform the following steps:

1. Sign in to the Azure portal: `https://portal.azure.com`.
2. In the search bar, type `virtual networks`; click on **Virtual networks** from the list of services shown.
3. From the **Virtual networks** blade, click on the **+ Create** option from the top menu of the blade, or use the **Create virtual network** button at the bottom of the blade. Set the **Project** and **Instance** details settings as required on the **Basics** tab.
4. No further configuration is required for this recipe. Click **Review + create**.
5. On the **Review + create** tab, click **Create**.
6. You will receive a notification that the deployment succeeded.

This *getting ready* task is complete.

Getting ready task – creating an Azure SQL database

Perform the following steps:

1. Sign in to the Azure portal: `https://portal.azure.com`.
2. Navigate to the **SQL databases** page or from the top menu of the *Azure Portal*; in the **Search** box, type `SQL databases`, and click on **SQL databases** from the results.
3. Click **+ Create** from the *top toolbar* from the **SQL databases** page.
4. From the **Basics** tab of the **Create SQL Database** page, under the **Project details** section, set your **Subscription** as required. Then, under **Resource group**, select **Create new**, enter a **Name**, and click **OK**.
5. Under the **Database details** section, enter a **Database name** as required, and for **Server**, click **Create new**.
6. From the **Create SQL Database Server** page, under the **Server details** section, set a **Server name** (*this must be globally unique*) and **Location** as required.
7. Under the **Authentication** section, set **Authentication method** to **Use SQL authentication**, and then enter a username and password as required. Then, click **OK**.
8. Set **Workload environment** to **Development** to reduce running costs for this recipe.

9. No additional configuration is required for this recipe; click **Review + Create**.

10. Click **Create** on the **Review + create** tab.

11. A notification will display that the resource deployment succeeded.

This *getting ready* task is complete. You are now ready to continue the main tasks for this recipe of implementing a private endpoint.

How to do it...

This recipe consists of the following tasks:

*   Creating an Azure SQL private endpoint

*   Cleaning up resources

Task – creating an Azure SQL private endpoint

Perform the following steps:

1. From the created **Azure SQL database** page, click **Configure** under **Configure access** from the **Getting started** page:

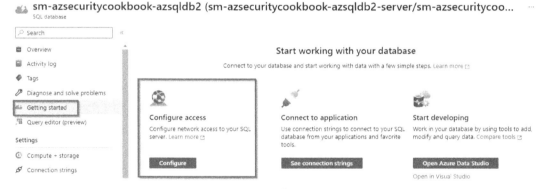

Figure 5.17 – Configuring access

2. From the **Networking** page, ensure that, from the **Public access** tab, **Public network access** is set to **Disable**:

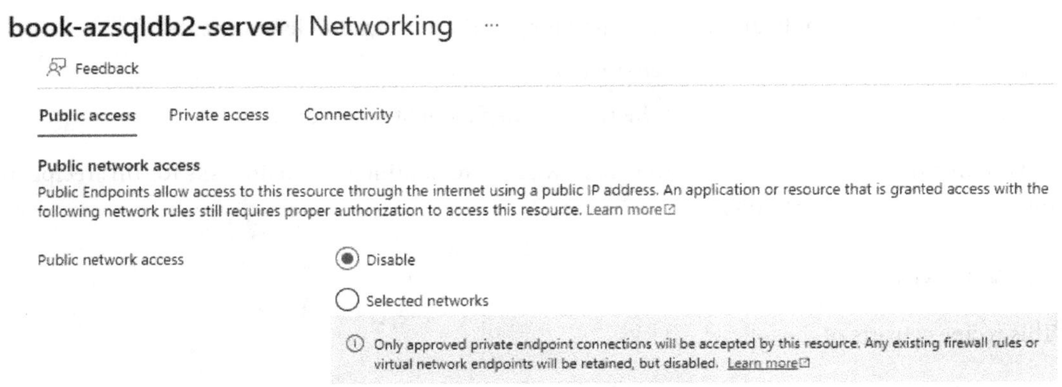

book-azsqldb2-server | Networking ···

🗩 Feedback

Public access   Private access   Connectivity

Public network access

Public Endpoints allow access to this resource through the internet using a public IP address. An application or resource that is granted access with the following network rules still requires proper authorization to access this resource. Learn more↗

Public network access   ◉ Disable

                                               ○ Selected networks

ⓘ Only approved private endpoint connections will be accepted by this resource. Any existing firewall rules or virtual network endpoints will be retained, but disabled. Learn more↗

Figure 5.18 – Disabling public network access

3. Navigate to the **Private access** tab and, under **Private endpoint connections**, click **Create a private endpoint**:

Figure 5.19 – Create a private endpoint

4. From the **Basics** tab of the **Create a private endpoint** page, set the **Project Details** and **Instance details** properties as required:

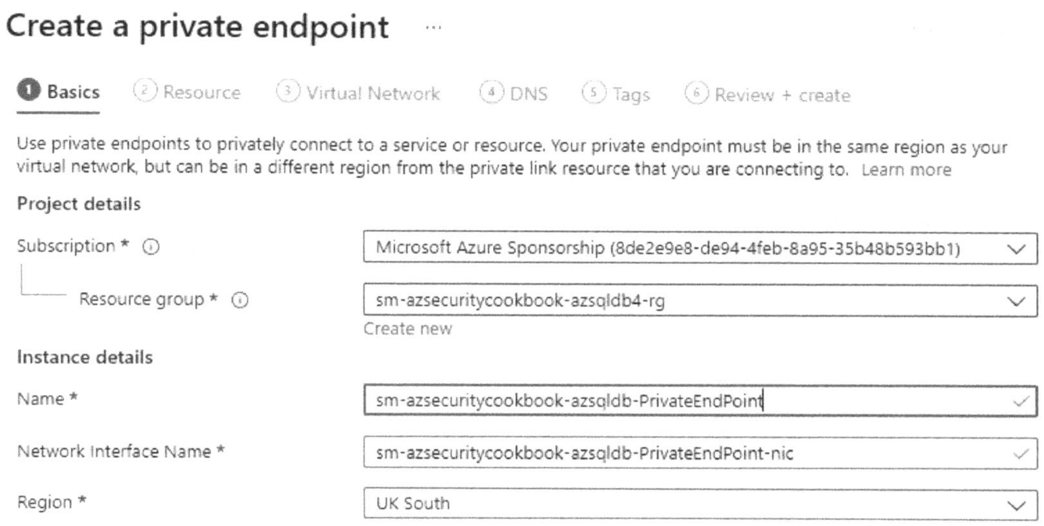

Create a private endpoint ...

① Basics   ② Resource   ③ Virtual Network   ④ DNS   ⑤ Tags   ⑥ Review + create

Use private endpoints to privately connect to a service or resource. Your private endpoint must be in the same region as your virtual network, but can be in a different region from the private link resource that you are connecting to. Learn more

Project details

Subscription * ⓘ

> Microsoft Azure Sponsorship (8de2e9e8-de94-4feb-8a95-35b48b593bb1)   ∨

Resource group * ⓘ

> sm-azsecuritycookbook-azsqldb4-rg   ∨

Create new

Instance details

Name *

> sm-azsecuritycookbook-azsqldb-PrivateEndPoint   ∨

Network Interface Name *

> sm-azsecuritycookbook-azsqldb-PrivateEndPoint-nic   ∨

Region *

> UK South   ∨

Figure 5.20 – Setting information for the private endpoint

5.   Click **Next : Resource**, and then click **Next: Virtual Network**.

6.   From the **Virtual Network** tab, ensure that you have selected the **Virtual network** and **Subnet** properties you want to deploy the private endpoint:

✓ Basics   ✓ Resource   **③ Virtual Network**   ④ DNS   ⑤ Tags   ⑥ Review + create

Networking

To deploy the private endpoint, select a virtual network subnet. Learn more

Virtual network * ⓘ

> sm-azsecuritycookbook-azsqldb4-vnet   ∨

Subnet * ⓘ

> sm-azsecuritycookbook-azsqldb4-vnet/default (10.1.0.0/24)   ∨

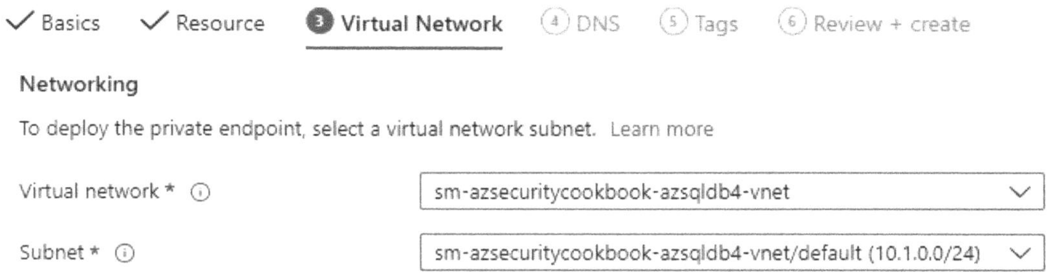

Figure 5.21 – Setting up a virtual network

7.   Click **Next: DNS**, click **Next: Tags**, and then click **Review + create**.

8.   From the **Review + create** tab, click **Create**.

With that, you have implemented a private endpoint. In the next task, we will clean up the resources that were created in this recipe.

Task – cleaning up resources

Perform the following steps:

1. From the **Search box** area in the Azure portal, type `resource groups` and select **Resource Groups** from the listed **Services** results.

2. From the **Resource groups** page, select the *resource group* we created for this recipe and click **Delete resource group**; this will delete all the resources that were created as part of this recipe:

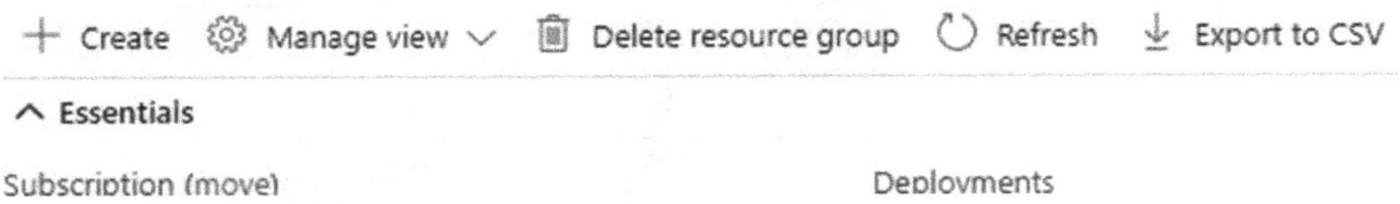

+ Create   ⚙ Manage view ⌄   🗑 Delete resource group   ↻ Refresh   ⤓ Export to CSV

⌃ Essentials

Subscription (move)   Deployments

Figure 5.22 – Delete resource group

With that, you have cleaned up the resources that were created in this recipe.

How it works….

In this recipe, we looked at implementing Azure Private Link to create a private endpoint that can connect to our PaaS services from an Azure virtual network. The benefit is that we do not expose our PaaS services to the public internet, and traffic can remain private over the Microsoft backbone.

As a getting ready task, we created an Azure Virtual Network Azure SQL database to illustrate the public PaaS service we want to connect to privately.

We wanted to demonstrate the ability to configure a public PaaS service to ensure that the connection traffic stays on the Microsoft backbone so that it bypasses the public internet.

There's more…

Now that we understand the concepts of private connectivity for public PaaS services such as Azure SQL database, we can explore this capability further. In the following Microsoft Learn article, you will learn how to privately connect a PaaS web application to a database's private endpoint; this ensures traffic only passes via the virtual network to the database and never over the internet:

*   Web app private connectivity to Azure SQL Database: `https://learn.microsoft.com/en-us/azure/architecture/example-scenario/private-web-app/private-web-app`

See also

Should you require further information, you can refer to the following Microsoft Learn articles:

- An overview of Azure SQL Database and SQL Managed Instance security capabilities: `https://learn.microsoft.com/en-us/azure/azure-sql/database/security-overview`

- What is Azure Private Link?: `https://learn.microsoft.com/en-us/azure/private-link/private-link-overview`

- Azure Private Link for Azure SQL Database and Azure Synapse Analytics: `https://learn.microsoft.com/en-us/azure/azure-sql/database/private-endpoint-overview`

- Azure SQL Database and Azure Synapse Analytics network access controls: `https://learn.microsoft.com/en-us/azure/azure-sql/database/network-access-controls-overview`

Implementing Azure AD authentication and authorization

With cloud-based implementations of databases such as Azure SQL, we can centrally manage, control, and protect the identities of users who access the databases when we use the **Azure Active Directory (Azure AD)** Cloud Identity Provider service.

Microsoft provides **Azure AD** as a fully managed **Identity Provider (IDP)** platform provided as **Software-as-a-Service (SaaS)**. Its primary function is to manage and control resource access through Authentication and Authorization.

Azure AD provides a mechanism to centrally authenticate users/groups for admin access to an Azure SQL database, without requiring local database accounts.

This recipe will teach you how to implement a centralized, controlled, and secure access management method for connecting to your Azure SQL databases using Azure AD authentication in place of local database accounts.

Getting ready

This recipe requires the following:

- A device with a browser, such as Edge or Chrome, to access the Azure portal: `https://portal.azure.com`

- Access to an Azure subscription, where you have access to the **Owner** role for the **Azure subscription**

- Access to an *Azure AD account* that has a **Global Administrator** role
- Access to an **Azure SQL database**; we will step through this process in the following *Getting ready* tasks

Continue with the following getting ready tasks for this recipe:

- Creating an Azure SQL database

Getting ready task – creating an Azure SQL database

Perform the following steps:

1. Sign in to the Azure portal: `https://portal.azure.com`.
2. Navigate to the **SQL databases** page or from the top menu of the Azure portal; in the **Search** box, type `SQL databases`, and click on **SQL databases** from the results.
3. Click **+ Create** from the *top toolbar* of the **SQL databases** page.
4. From the **Basics** tab of the **Create SQL Database** page, under the **Project details** section, set a **Subscription** as required. Then, under **Resource group**, select **Create new**, enter a **Name**, and click **OK**.
5. Under the **Database details** section, enter a **Database name** as required, and for **Server**, click **Create new**.
6. From the **Create SQL Database Server** page, under the **Server details** section, set a **Server name** (*this must be globally unique*) and **Location** as required.
7. Under the **Authentication** section, set **Authentication method** to **Use SQL authentication** and enter a username and password as required. Then, click **OK**.
8. Set **Workload environment** to **Development** to reduce running costs for this recipe.
9. No additional configuration is required for this recipe; click **Review + Create**.
10. Click **Create** on the **Review + create** tab.
11. A notification will display that the resource deployment succeeded.

This *getting ready* task is complete. You are now ready to configure Azure AD authentication for accessing Azure SQL.

How to do it...

This recipe consists of the following tasks:

- Configuring authentication to Azure AD for Azure SQL databases
- Cleaning up resources

Task – configuring authentication to Azure AD for Azure SQL databases

Perform the following steps:

1. Sign in to the Azure portal: `https://portal.azure.com`.

2. Navigate to the **SQL Servers** page or from the top menu of the Azure portal; in the **Search** box, type `SQL server`, and click on **SQL servers** from the results.

3. Open the page for your created **SQL server** and click **Azure Active Directory** under the **Settings** section:

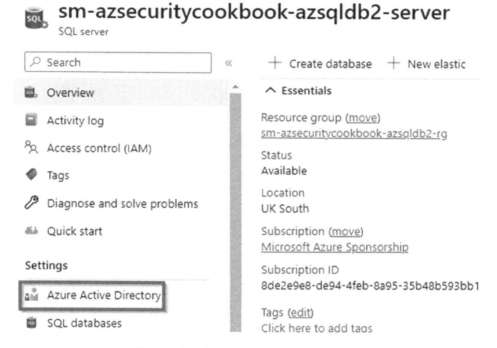

Figure 5.23 – Azure Active Directory settings

4. From the **Azure Active Directory** page, click **Set admin** from the *top toolbar*:

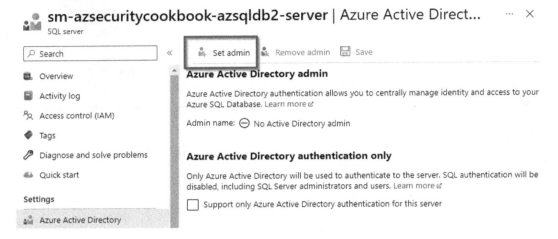

Figure 5.24 – Setting an Azure Active Directory admin

5. Search for a user identity as required from your directory:

Figure 5.25 – Selecting a user

6. Click the user identity and click **Select**.

7. From the **Azure Active Directory** page, under **Azure Active Directory authentication section only,** check the **Support only Azure Active Directory authentication to this server** box.

8. From the *pop-up* dialog box, click **Yes**:

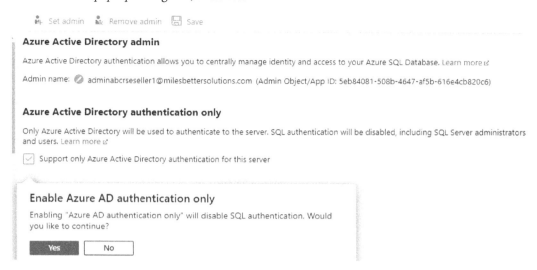

Figure 5.26 – Enable Azure AD authentication only

9. From the top toolbar, click **Save**:

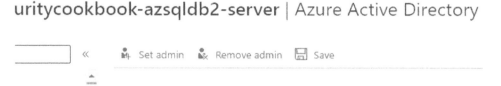

Figure 5.27 – Saving your settings

This task is completed. In the next task, we will clean up the resources that were created in this recipe.

Task – cleaning up resources

Perform the following steps:

1. From the **Search box** area in the Azure portal, type `resource  groups` and select **Resource Groups** from the listed **Services** results.

2. From the **Resource groups** page, select the *resource group* we created for this recipe and click **Delete resource group**; this will delete all the resources that were created as part of this recipe:

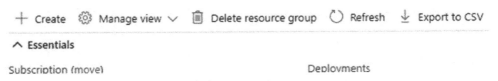

Figure 5.28 – Delete resource group

With that, you have cleaned up the resources that were created in this recipe.

How it works...

In this recipe, we looked at centrally managing the identity of database users using Azure AD authentication and removing access from local SQL authentication accounts.

As a getting ready task, we created an Azure SQL database to configure the Azure AD authentication mechanism.

See also

Should you require further information, you can refer to the following Microsoft Learn articles:

- An overview of Azure SQL Database and SQL Managed Instance security capabilities: https://learn.microsoft.com/en-us/azure/azure-sql/database/security-overview

- Use Azure Active Directory authentication: https://learn.microsoft.com/en-us/azure/azure-sql/database/authentication-aad-overview

- Exercise – Control who can access your database: https://learn.microsoft.com/en-us/training/modules/secure-your-azure-sql-database/3-manage-authentication

Securing Azure Storage

In the previous chapter, we covered recipes for securing **Azure databases**.

We introduced the concept of the *shared responsibility model* for public cloud platform service providers. The same approach and model applies to this chapter; the customer is responsible for the security of storage held in Azure and enabling and configuring the appropriate level of protection and security controls for the customer's needs.

This chapter concludes *Part 1* of this cookbook, in which we have been looking at Azure's security features and capabilities.

In this final chapter of this part, the recipes we look at will cover the security settings that can be configured and data protection through encryption.

By the end of this chapter, you will have learned the following skills to carry out the following recipes to secure Azure Storage:

- Implementing security settings on storage accounts
- Implementing network security
- Implementing encryption

Technical requirements

For this chapter, it is already assumed that you have an *Azure AD tenancy* and an *Azure subscription* from completing the recipes in previous chapters of this cookbook. If you skipped straight to this section, the information to create a new *Azure AD tenancy* and an *Azure subscription* for these recipes is included in the following list of requirements.

For this chapter, the following are required:

- A device with a browser, such as *Edge* or *Chrome*, to access the Azure portal at `https://portal.azure.com`.

- An **Azure AD tenancy** and **Azure subscription**; you may use existing ones or sign up for free: `https://azure.microsoft.com/en-us/free`.

- An **Owner role** for the **Azure subscription**.

Implementing security settings on storage accounts

Azure Storage accounts, by default, have an *internet-accessible public endpoint*, the same as we learned about with Azure SQL databases from the last chapter. Therefore, we must provide security and access control layers for our *defense-in-depth strategy*.

This recipe will teach you to secure and control access to your Azure storage accounts.

We will look at the security settings that can be configured when creating storage accounts, network security, and encryption.

Getting ready

This recipe requires the following:

- A device with a browser, such as *Edge* or *Chrome*, to access the **Azure portal**: `https://portal.azure.com`

- Access to an **Azure subscription**, where you have access to the **Owner** role for the **Azure subscription**

How to do it...

This task consists of the following tasks:

- Reviewing the security settings when creating a storage account

Task – reviewing the security settings when creating a storage account

Perform the following steps:

1. Sign in to the Azure portal at `https://portal.azure.com`.

2. In the search bar, type `storage accounts`; click on **Storage accounts** from the list of services shown:

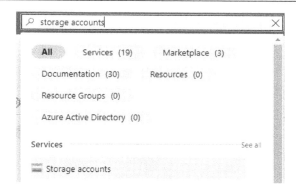

Figure 6.1 – Searching for Storage accounts

3.   On the **Storage accounts** blade, click on the **+ Create** option from the top menu, or use the **Create storage account** button at the bottom of the blade.

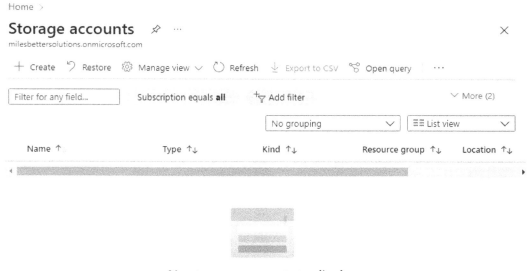

Figure 6.2 – Create a storage account

4.   Set the **Project** and **Instance** details settings as required on the **Basics** tab:

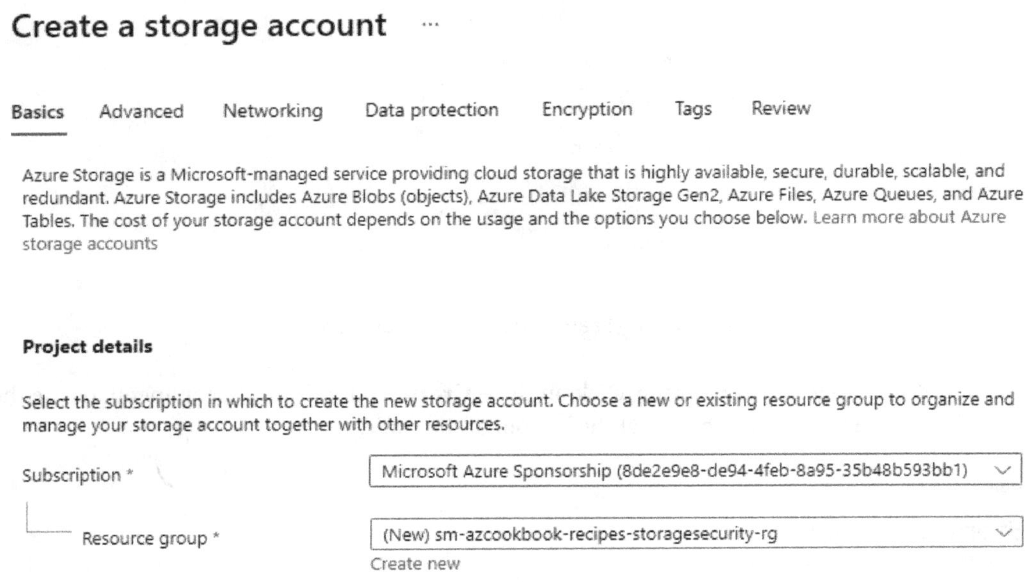

Create a storage account   ⋯

Basics   Advanced   Networking   Data protection   Encryption   Tags   Review

Azure Storage is a Microsoft-managed service providing cloud storage that is highly available, secure, durable, scalable, and redundant. Azure Storage includes Azure Blobs (objects), Azure Data Lake Storage Gen2, Azure Files, Azure Queues, and Azure Tables. The cost of your storage account depends on the usage and the options you choose below. Learn more about Azure storage accounts

Project details

Select the subscription in which to create the new storage account. Choose a new or existing resource group to organize and manage your storage account together with other resources.

Subscription *   Microsoft Azure Sponsorship (8de2e9e8-de94-4feb-8a95-35b48b593bb1)   ⌄

Resource group *   (New) sm-azcookbook-recipes-storagesecurity-rg   ⌄
                      Create new

Figure 6.3 – Setting the project details

5.   Under **Instance details**, set the **Storage account name** and **Region** details as required:

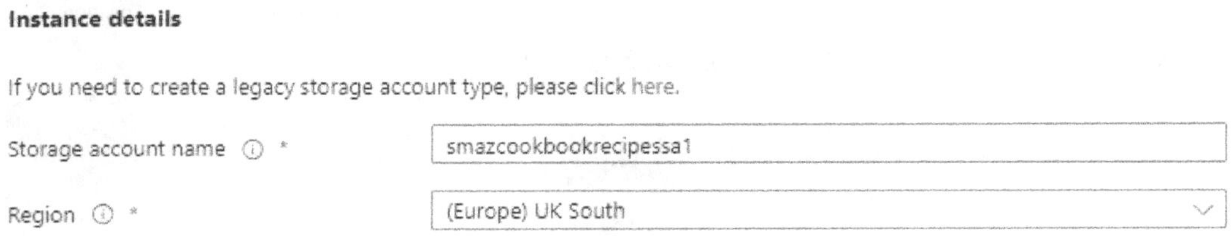

Instance details

If you need to create a legacy storage account type, please click here.

Storage account name ⓘ *   smazcookbookrecipessa1

Region ⓘ *   (Europe) UK South   ⌄

Figure 6.4 – Setting the instance details

6.   Leave the **Performance** and **Redundancy** settings at their defaults for this recipe.

7.   Click on **Next : Advanced**.

8.   On the **Advanced** tab, in the **Security** section, we can now review the *security settings* configured at creation time:

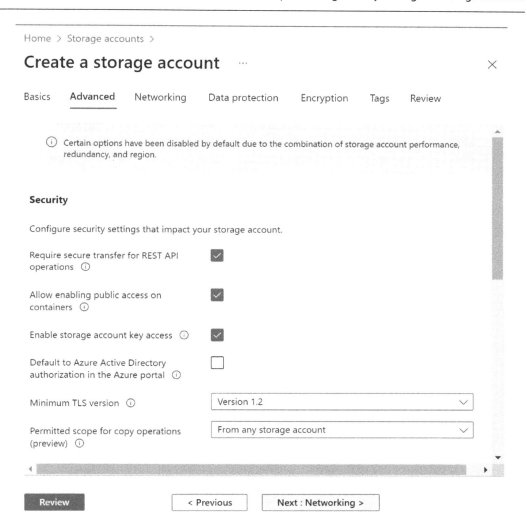

Figure 6.5 – Security settings for the storage account

- **Require secure transfer for REST API operations**: This setting ensures that only **HTTPS** requests can be made to the storage account. Learn more here: `https://learn.microsoft.com/en-us/azure/storage/common/storage-require-secure-transfer`.

- **Allow enabling public access on containers**: This setting permits *anonymous public endpoint access* to a storage account **container**. **Unchecking** this checkbox **disables** *anonymous public endpoint access* to a storage account container. Learn more here: `https://learn.microsoft.com/en-us/azure/storage/blobs/anonymous-read-access-prevent`.

- **Enable storage account key access**: This setting allows authorization of access to the storage account using a **shared key** or **shared access signatures (SAS)**. **Unchecking** this checkbox **disables** and **denies** *key access*. Learn more here: `https://learn.microsoft.com/ en-us/azure/storage/common/shared-key-authorization-prevent`.

- **Default to Azure Active Directory authorization in the Azure portal**: This setting allows the authorization of access to the storage account to a default of using **identity-based authentication**, such as **Azure AD**. Learn more here: `https://learn.microsoft. com/en-us/azure/storage/blobs/authorize-data-operations- portal#default-to-azure-ad-authorization-in-the-azure-portal`.

9. **Minimum TLS version**: This setting allows you to specify the default minimum **TLS version**. When set to the default of **Version 1.2**, requests are rejected when made using **TLS 1.0** or **TLS 1.1**.

    Learn more here:

    `https://learn.microsoft.com/en-us/azure/storage/common/transport- layer-security-configure-minimum-version`

10. **Permitted scope for copy operations**: This setting implements the limits of the copy operations for lateral movement and data breach.

11. No further configuration is required for this recipe.

12. Click **Review**.

13. On the **Review** tab, click **Create**. You will receive a notification that the deployment succeeded.

14. Click on **Go to resource** ready for the next task for this recipe.

This task of reviewing the security settings that can be implemented when creating a storage account is complete. In the next task, we will clean up the resources created in this recipe.

Task – cleaning up resources

Perform the following steps:

1. In the search box in the *Azure portal*, type `resource groups` and select **Resource Groups** from the listed **Services** results.

2. On the **Resource groups** page, select the *resource group* we created for this recipe, and click **Delete resource group**; this will delete all the resources created as part of this recipe.

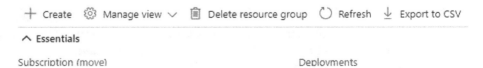

Figure 6.6 – Delete resource group

This task of cleaning up the resources created in this recipe is complete.

How it works...

For this recipe, we looked at reviewing the security setting options available when creating a storage account. We saw the default security settings that are applied and the optional security settings that can be applied.

See also

Should you require further information, you can refer to the following Microsoft Learn articles:

- *Implement storage security*: `https://learn.microsoft.com/en-us/training/modules/storage-security`

- *Create a storage account*: `https://learn.microsoft.com/en-gb/azure/storage/common/storage-account-create`

- *Storage account overview*: `https://learn.microsoft.com/en-us/azure/storage/common/storage-account-overview`

- *Azure Storage redundancy*: `https://learn.microsoft.com/en-us/azure/storage/common/storage-redundancy`

Implementing network security

We must secure not only the storage account itself but also the *network* we use for access; this enforces our *defense-in-depth* strategy.

This recipe will teach you to secure *network access* to your storage accounts.

We will look at the network access settings that can be configured when creating storage accounts, virtual network access, and implementing a storage IP firewall.

This recipe requires the following:

- A device with a browser, such as *Edge* or *Chrome*, to access the **Azure portal** at `https://portal.azure.com`.

- Access to an **Azure subscription**, where you have access to the **Owner** role for the **Azure subscription**.

How to do it...

This task consists of the following tasks:

- Implementing network access settings when creating a storage account
- Implementing virtual network access
- Implementing a storage IP firewall

Task – implementing network access settings when creating a storage account

Perform the following steps:

1. Sign in to the Azure portal at `https://portal.azure.com`.

2. In the search bar, type `storage accounts`; click on **Storage accounts** from the list of services shown.

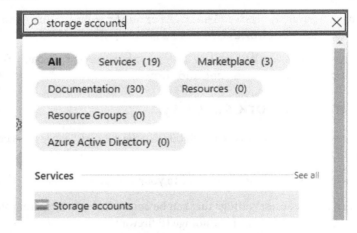

Figure 6.7 – Searching for storage accounts

3. On the **Storage accounts** blade, click on the **+ Create** option from the top menu, or use the **Create storage account** button at the bottom of the blade.

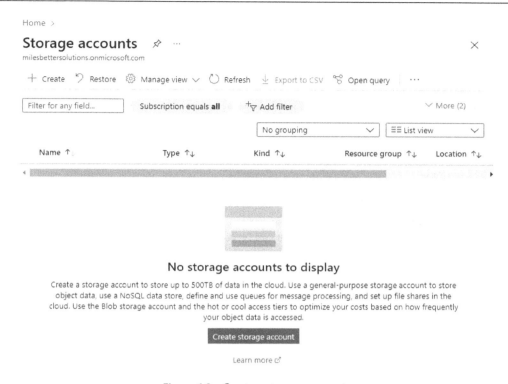

Figure 6.8 – Create a storage account

4.   Set the **Project** and **Instance** details settings as required on the **Basics** tab.

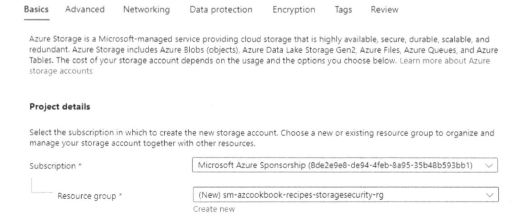

Figure 6.9 – Setting the project details

5.   Under **Instance details**, set the **Storage account name** and **Region** details as required.

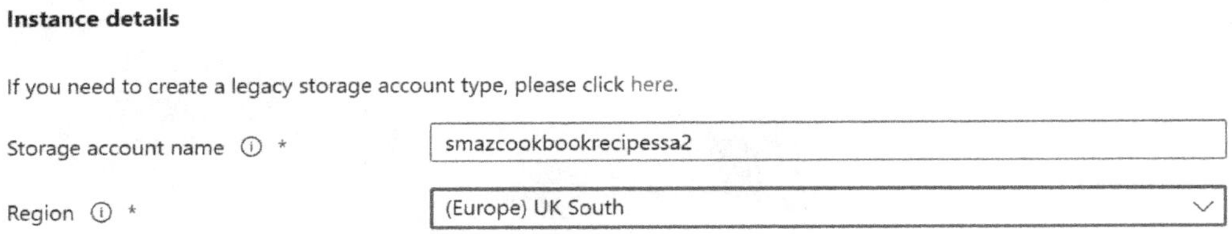

Instance details

If you need to create a legacy storage account type, please click here.

Storage account name ⓘ *   smazcookbookrecipessa2

Region ⓘ *   (Europe) UK South   ⌄

Figure 6.10 – Set instance details

6.   Click **Next : Advanced** and click **Next : Networking**.

7.   On the **Networking** tab, under the **Network connectivity** section, we can now review the following **Network access** settings that can be configured at the time of creation.

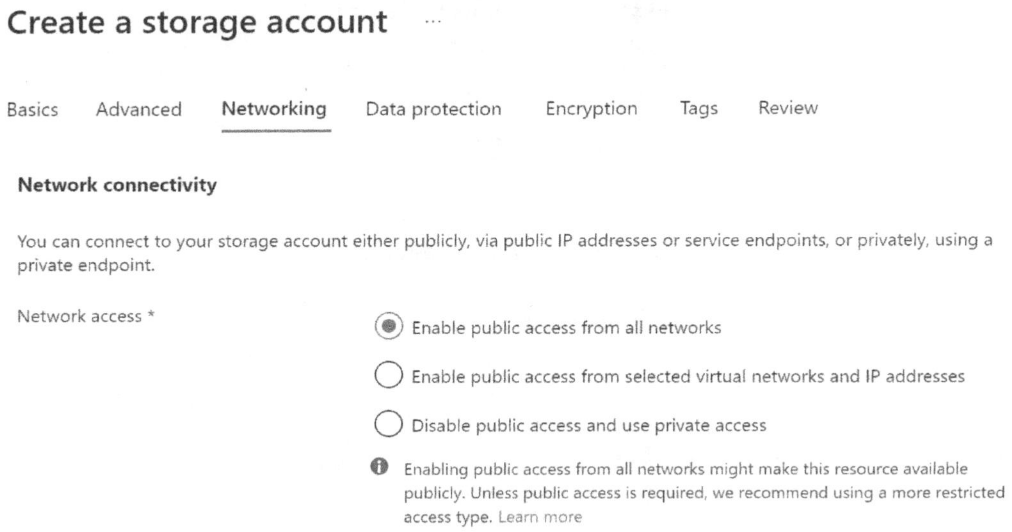

Create a storage account   ⋯

Basics   Advanced   **Networking**   Data protection   Encryption   Tags   Review

Network connectivity

You can connect to your storage account either publicly, via public IP addresses or service endpoints, or privately, using a private endpoint.

Network access *   ⦿ Enable public access from all networks

                          ◯ Enable public access from selected virtual networks and IP addresses

                          ◯ Disable public access and use private access

                          ❶ Enabling public access from all networks might make this resource available
                             publicly. Unless public access is required, we recommend using a more restricted
                             access type. Learn more

Figure 6.11 – Network connectivity settings for the storage account

• **Enable public access from all networks**: This setting is the default; the storage account's public endpoint will allow traffic to be routed from all networks. Consider the security implications of this "*any network*" public access.

• **Enable public access from selected virtual networks and IP addresses**: This setting requires all access to the storage account to be routed via a virtual network; only the selected virtual network will be able to access the storage account.

- **Disable public access and use private access**: This setting will remove public access. Access to the storage account can only be provided through a private connection; a private endpoint must be created.

For this recipe, we will leave the setting as the *default* of **Enable public access from all networks**.

8. In the **Network routing** section, you can determine the **Routing preference** setting; there are two options:

- **Microsoft network routing**: This setting will direct traffic to the Microsoft backbone (private) network as close to the source as possible

- **Internet routing**: This setting will direct traffic to enter the Microsoft backbone (private) network closer to the Azure endpoint

For this recipe, we will leave the setting as the *default* of **Enable public access from all networks**.

9. Click **Review**.

10. On the **Review** tab, click **Create**.

11. You will receive a notification that the deployment succeeded.

12. Click on **Go to resource** ready for the next task for this recipe.

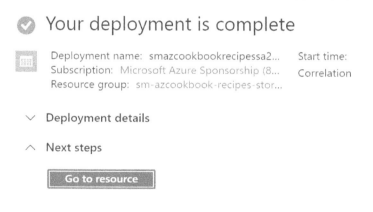

Figure 6.12 – Resource deployment complete

This task of reviewing the network access settings that can be implemented when creating a storage account is complete. In the next task, we will look at implementing virtual network access options.

Task – implementing virtual network access

Perform the following steps:

1. On the created *storage account* page, click on **Networking** under the **Security + networking** section from the left toolbar.

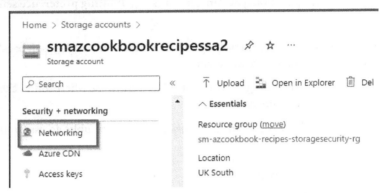

Figure 6.13 – Storage account page

2. On the **Firewalls and virtual networks** tab on the **Networking** page, select the **Enabled from selected virtual network and IP addresses** option in the **Public network access** section.

Figure 6.14 – Public network settings

3. In the **Virtual networks** section, you can select **Add existing virtual network**.

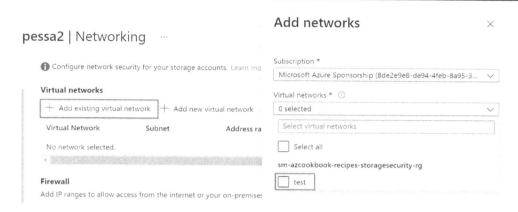

Figure 6.15 – Add existing virtual network

4. Alternatively, you can select **Add new virtual network**.

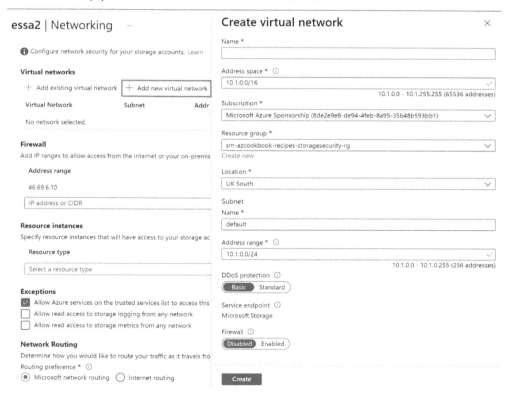

Figure 6.16 – Add new virtual network

This task is complete. In the next task, we will implement a storage IP firewall.

Task – implementing a storage IP firewall

Perform the following steps:

1. On the created *storage account* page, click **Networking** in the **Security + networking** section from the left toolbar.

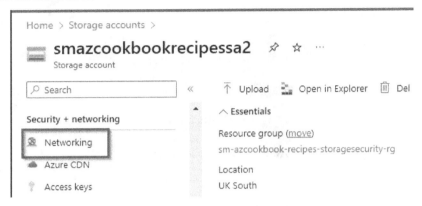

Figure 6.17 – Storage account page

2. On the **Firewalls and virtual networks** tab on the **Networking** page, select the **Enabled from selected virtual network and IP addresses** option in the **Public network access** section:

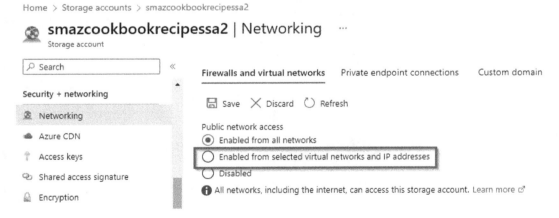

Figure 6.18 – Public network settings

3. In the **Firewall** section, check the **Add your client IP address ('Your-IP')** checkbox.

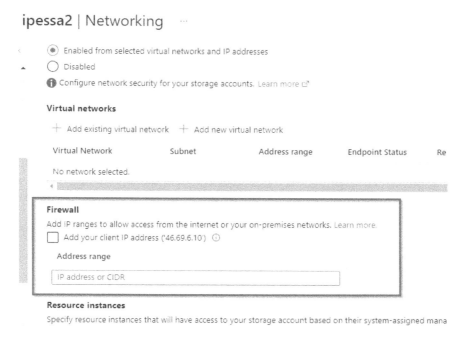

Figure 6.19 – Storage firewall settings

4. In **Address range**, add any *IP address* or *CIDR IP block* that you wish to allow to access the *storage account*.

5. Click **Save** from the top toolbar of the **Firewalls and virtual networks** tab.

This task of setting a storage firewall rule is complete. In the next task, we will clean up the resources created in this recipe.

Task – cleaning up resources

Perform the following steps:

1. In the search box in the *Azure portal*, type `resource groups` and select **Resource Groups** from the listed **Services** results.

2. On the **Resource groups** page, select the *resource group* we created for this recipe, and click **Delete resource group**; this will delete all the resources created as part of this recipe.

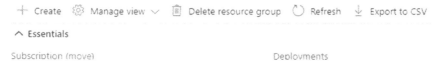

Figure 6.20 – Delete resource group

This task of cleaning up the resources created in this recipe is complete.

How it works...

For this recipe, we looked at implementing the available network security capabilities for a storage account. We saw the security settings that can be set when creating a storage account. We looked at the options to restrict access to the storage account from selected virtual networks. We concluded with how to implement a storage IP firewall to restrict access by IP address or CIDR address block.

There's more...

This recipe covered tasks including controlling network access through virtual networks and a storage IP firewall. In addition, you can also create private endpoints, so access via the internet is not possible and connections are only possible via a virtual network. A private endpoint works by assigning the storage account a network interface and private IP address for your virtual network's private IP address space; this allows the service to be brought into the virtual network for secure private access.

The **Private endpoint connections** page for a storage account is represented in the following figure.

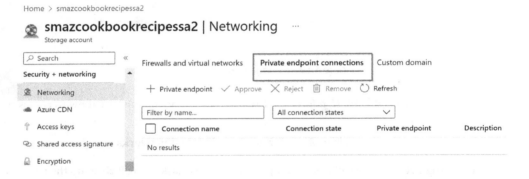

Figure 6.21 – Storage private endpoint connections

See also

Should you require further information, you can refer to the following Microsoft Learn articles:

- What is a private endpoint?: `https://learn.microsoft.com/en-gb/azure/storage/common/storage-network-security`

- *Configure Azure Storage firewalls and virtual networks*: `https://learn.microsoft.com/en-gb/azure/storage/common/storage-network-security`

- *Network routing preference for Azure Storage*: `https://learn.microsoft.com/en-gb/azure/storage/common/network-routing-preference`

- *Virtual Network service endpoints*: `https://learn.microsoft.com/en-us/azure/virtual-network/virtual-network-service-endpoints-overview`

Implementing encryption

When implementing a **Zero Trust** approach to cloud resource security, we must adopt the stance of **Assumed Breach**. This means assuming that *bad actors* have already *compromised* our *perimeter defenses,* and we are at risk of *data integrity breaches* and *data exfiltration*. Therefore, we must provide mechanisms that ensure our data's integrity remains and our data is unreadable and unusable in the case of exfiltration.

This recipe will teach you to secure your storage account data at rest through encryption using **Azure Storage Service Encryption**.

In this task, we will look at the **customer-managed keys** encryption type; by default, storage accounts are encrypted by **Microsoft-managed keys** with no configuration required.

We will look at how encryption can be set for existing storage accounts.

Getting ready

This recipe requires the following:

- A device with a browser, such as *Edge* or *Chrome*, to access the **Azure portal** at `https://portal.azure.com`

- Access to an **Azure subscription**, where you have access to the **Owner** role for the **Azure subscription**

- Access to an **Azure storage account**; we will step through this process in the following *Getting ready* tasks

Continue with the following *Getting ready* tasks for this recipe:

- Creating an Azure storage account

Getting ready task – creating an Azure storage account

Perform the following steps:

1. Sign in to the Azure portal at `https://portal.azure.com`.

2. In the search bar, type `storage accounts`; click on **Storage accounts** from the list of services shown.

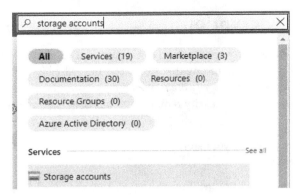

Figure 6.22 – Searching for Storage accounts

3.   On the **Storage accounts** blade, click on the **+ Create** option from the top menu, or use the **Create storage account** button at the bottom of the blade.

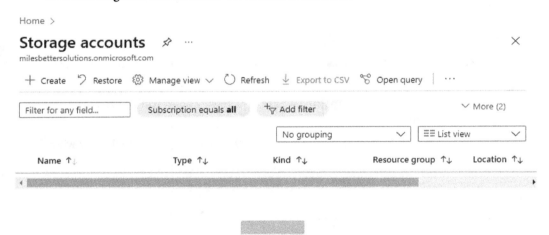

No storage accounts to display

Create a storage account to store up to 500TB of data in the cloud. Use a general-purpose storage account to store object data, use a NoSQL data store, define and use queues for message processing, and set up file shares in the cloud. Use the Blob storage account and the hot or cool access tiers to optimize your costs based on how frequently your object data is accessed.

Learn more ⬚

Figure 6.23 – Create storage account

4. Set the **Project** and **Instance** details settings as required on the **Basics** tab.

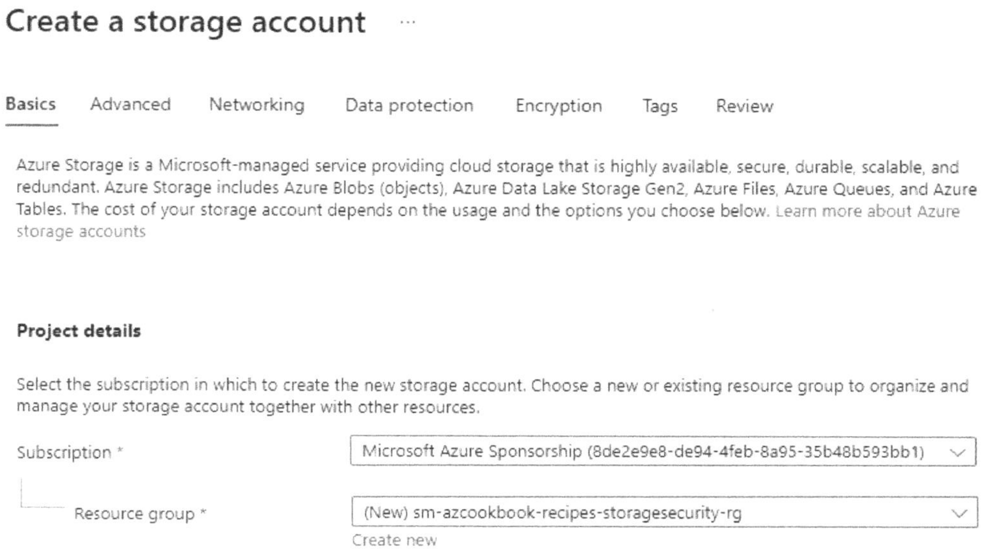

Create a storage account   ...

Basics   Advanced   Networking   Data protection   Encryption   Tags   Review

Azure Storage is a Microsoft-managed service providing cloud storage that is highly available, secure, durable, scalable, and redundant. Azure Storage includes Azure Blobs (objects), Azure Data Lake Storage Gen2, Azure Files, Azure Queues, and Azure Tables. The cost of your storage account depends on the usage and the options you choose below. Learn more about Azure storage accounts

Project details

Select the subscription in which to create the new storage account. Choose a new or existing resource group to organize and manage your storage account together with other resources.

Subscription *   Microsoft Azure Sponsorship (8de2e9e8-de94-4feb-8a95-35b48b593bb1)   ∨

Resource group *   (New) sm-azcookbook-recipes-storagesecurity-rg   ∨
                      Create new

Figure 6.24 – Setting the project details

5. Under **Instance details**, set the **Storage account name** and **Region** details as required.

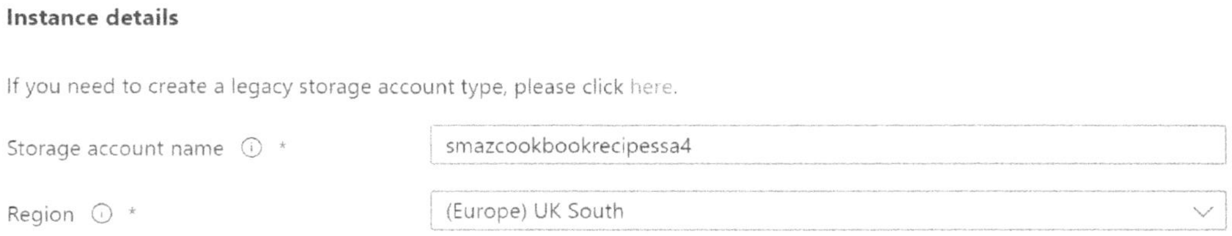

Instance details

If you need to create a legacy storage account type, please click here.

Storage account name ⓘ *   smazcookbookrecipessa4

Region ⓘ *   (Europe) UK South   ∨

Figure 6.25 – Setting the instance details

6. No further configuration is required for this recipe.

7. Click **Review**.

8. On the **Review** tab, click **Create**.

9. You will receive a notification that the deployment succeeded.

10. Click on **Go to resource** ready for the main task for this recipe.

This *Getting ready* task is complete. You are now ready to continue the main tasks for this recipe of setting encryption for an existing storage account.

How to do it...

This task consists of the following tasks:

- Implementing encryption for an existing storage account

Task – implementing encryption for an existing storage account

Perform the following steps:

1. On the created **Azure storage account** page, click on **Encryption** in the **Security + networking** section.

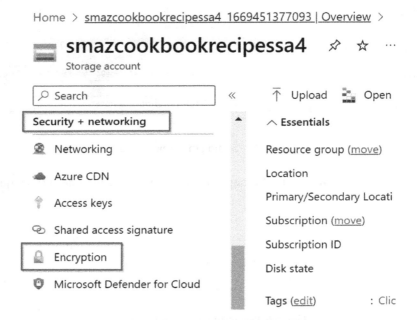

Figure 6.26 – Storage account page

2. On the **Encryption** page, select **Customer-managed keys** in the **Encryption selection** section of the **Encryption** tab.

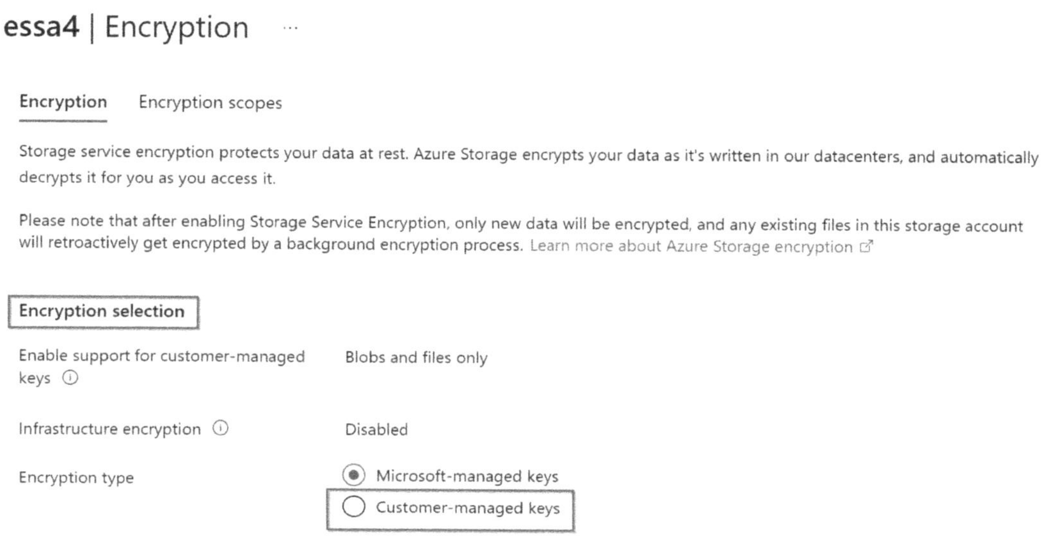

essa4 | Encryption   ...

Encryption   Encryption scopes

Storage service encryption protects your data at rest. Azure Storage encrypts your data as it's written in our datacenters, and automatically decrypts it for you as you access it.

Please note that after enabling Storage Service Encryption, only new data will be encrypted, and any existing files in this storage account will retroactively get encrypted by a background encryption process. Learn more about Azure Storage encryption ☐

| Encryption selection |

| Enable support for customer-managed keys ⓘ | Blobs and files only |

| Infrastructure encryption ⓘ | Disabled |

Encryption type
- ⦿ Microsoft-managed keys
- ◯ Customer-managed keys

Figure 6.27 – Encryption page

3. In the **Key selection** section, click **Select a key vault and key**.

Key selection

Encryption key
- ⦿ Select from key vault
- ◯ Enter key URI

Key vault and key * | Select a key vault and key

Figure 6.28 – Customer-managed keys

4. On the **Select a key** page, select **Key vault** for the **Key store type** option:

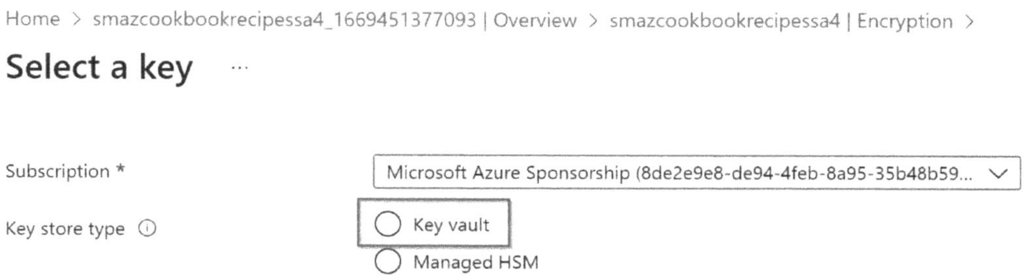

Home > smazcookbookrecipessa4_1669451377093 | Overview > smazcookbookrecipessa4 | Encryption >

Select a key   ...

Subscription *   Microsoft Azure Sponsorship (8de2e9e8-de94-4feb-8a95-35b48b59... ⌄

Key store type ⓘ
- ◯ Key vault
- ◯ Managed HSM

Figure 6.29 – Key store type

5.   For the **Key vault** selection, click on **Create new key vault**.

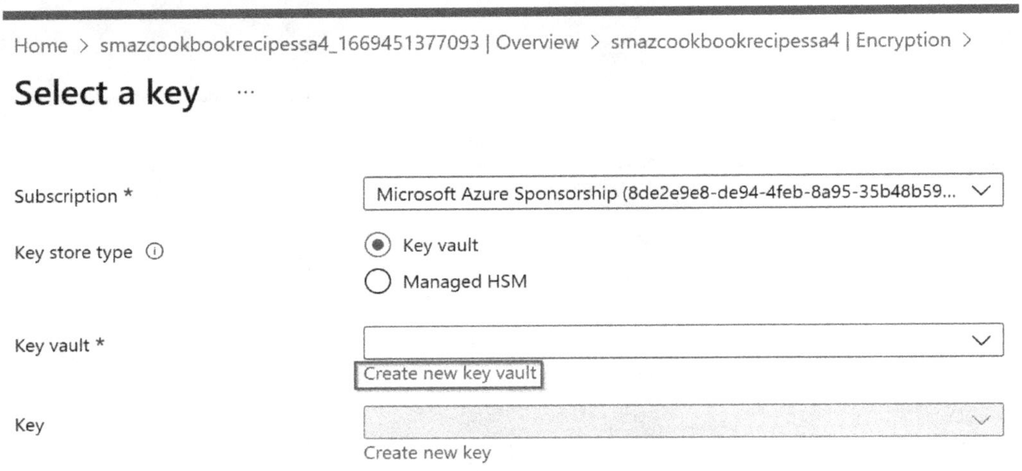

Home > smazcookbookrecipessa4_1669451377093 | Overview > smazcookbookrecipessa4 | Encryption >

Select a key   ...

Subscription *	Microsoft Azure Sponsorship (8de2e9e8-de94-4feb-8a95-35b48b59... ⌄
Key store type ⓘ	⦿ Key vault ◯ Managed HSM
Key vault *	⌄
	Create new key vault
Key	⌄
	Create new key

Figure 6.30 – Key vault selection

6.   On the **Create a key vault** page, enter the following:

- For **Project details**, select the same **Subscription** and **Resource group** details used to create the storage account in the *Getting ready* task for this recipe

- For **Instance details**, enter the **Key vault name** details as required, and select the same **Region** option used to create the storage account in the *Getting ready* task for this recipe

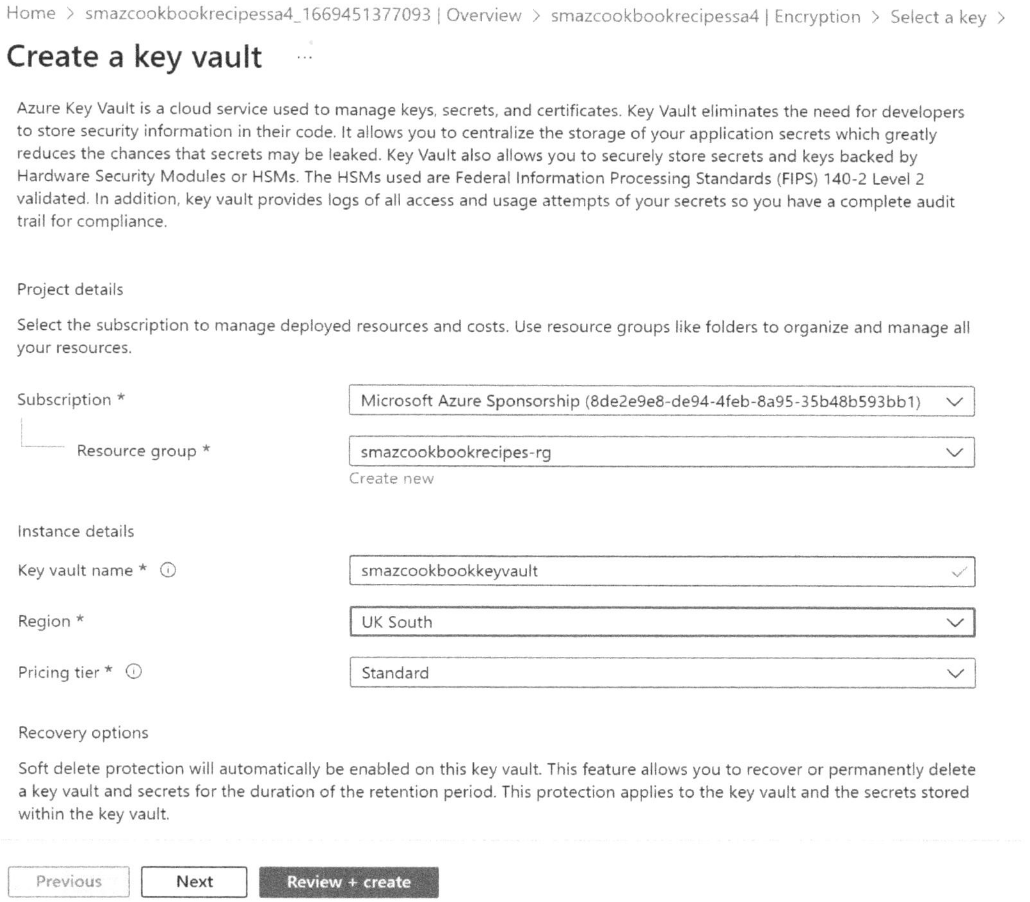

Home > smazcookbookrecipessa4_1669451377093 | Overview > smazcookbookrecipessa4 | Encryption > Select a key >

Create a key vault   ...

Azure Key Vault is a cloud service used to manage keys, secrets, and certificates. Key Vault eliminates the need for developers to store security information in their code. It allows you to centralize the storage of your application secrets which greatly reduces the chances that secrets may be leaked. Key Vault also allows you to securely store secrets and keys backed by Hardware Security Modules or HSMs. The HSMs used are Federal Information Processing Standards (FIPS) 140-2 Level 2 validated. In addition, key vault provides logs of all access and usage attempts of your secrets so you have a complete audit trail for compliance.

Project details

Select the subscription to manage deployed resources and costs. Use resource groups like folders to organize and manage all your resources.

Subscription * | Microsoft Azure Sponsorship (8de2e9e8-de94-4feb-8a95-35b48b593bb1)   ∨ |

└──── Resource group * | smazcookbookrecipes-rg   ∨ |
                                 Create new

Instance details

Key vault name * ⓘ | smazcookbookkeyvault   ✓ |

Region * | UK South   ∨ |

Pricing tier * ⓘ | Standard   ∨ |

Recovery options

Soft delete protection will automatically be enabled on this key vault. This feature allows you to recover or permanently delete a key vault and secrets for the duration of the retention period. This protection applies to the key vault and the secrets stored within the key vault.

[Previous]   [Next]   [Review + create]

Figure 6.31 – Key vault creation

No further configuration is required for this recipe.

7. Click **Review + create**.

8. On the **Review + create** tab, click **Create**.

9. You will receive a notification that the deployment succeeded; wait to be returned to the **Select a key** page.

10. Once you have returned to the **Select a key** page, click **Create a new key** for the **Key** value.

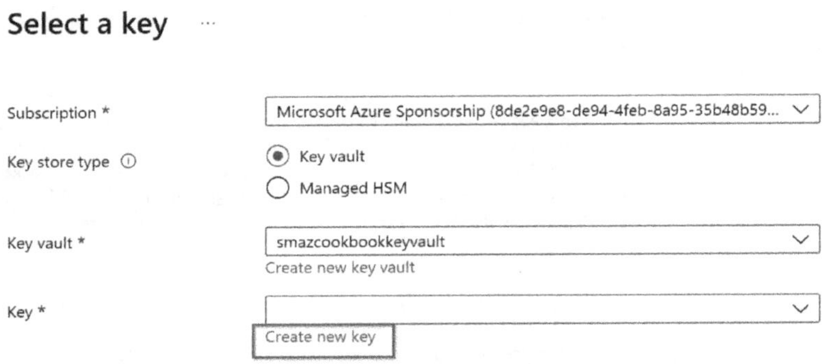

Select a key ...

Subscription *	Microsoft Azure Sponsorship (8de2e9e8-de94-4feb-8a95-35b48b59... ∨
Key store type ⓘ	⦿ Key vault ◯ Managed HSM
Key vault *	smazcookbookkeyvault ∨ Create new key vault
Key *	⌄ Create new key

Figure 6.32 – Key selection

11. On the **Create a key** page, enter a **Name** value as required, and then click **Create**.

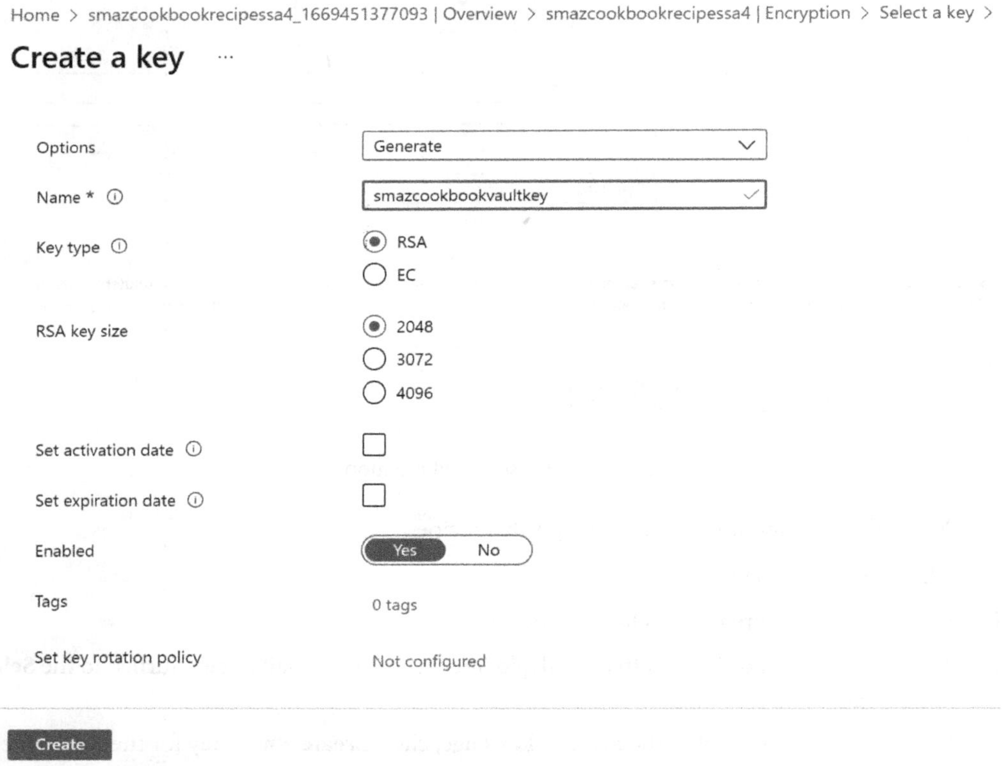

Home > smazcookbookrecipessa4_1669451377093 | Overview > smazcookbookrecipessa4 | Encryption > Select a key >

Create a key ...

Options	Generate ∨
Name * ⓘ	smazcookbookvaultkey ✓
Key type ⓘ	⦿ RSA ◯ EC
RSA key size	⦿ 2048 ◯ 3072 ◯ 4096
Set activation date ⓘ	☐
Set expiration date ⓘ	☐
Enabled	(Yes No)
Tags	0 tags
Set key rotation policy	Not configured

Create

Figure 6.33 – Creating a new key

12. Once you have returned to the **Select a key** page, click **Select**.

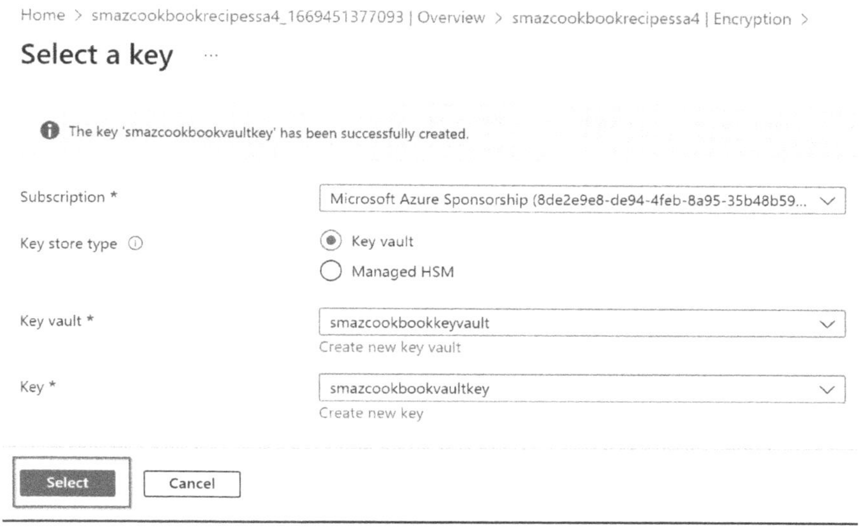

Home > smazcookbookrecipessa4_1669451377093 | Overview > smazcookbookrecipessa4 | Encryption >

Select a key   ⋯

ⓘ The key 'smazcookbookvaultkey' has been successfully created.

Subscription * | Microsoft Azure Sponsorship (8de2e9e8-de94-4feb-8a95-35b48b59... ∨ |

Key store type ⓘ   ◉ Key vault
                                   ○ Managed HSM

Key vault * | smazcookbookkeyvault   ∨ |
                                   Create new key vault

Key * | smazcookbookvaultkey   ∨ |
                                   Create new key

[Select]   [Cancel]

Figure 6.34 – Select a key

13. Once you have returned to the main **Encryption** settings page, click **Save**.

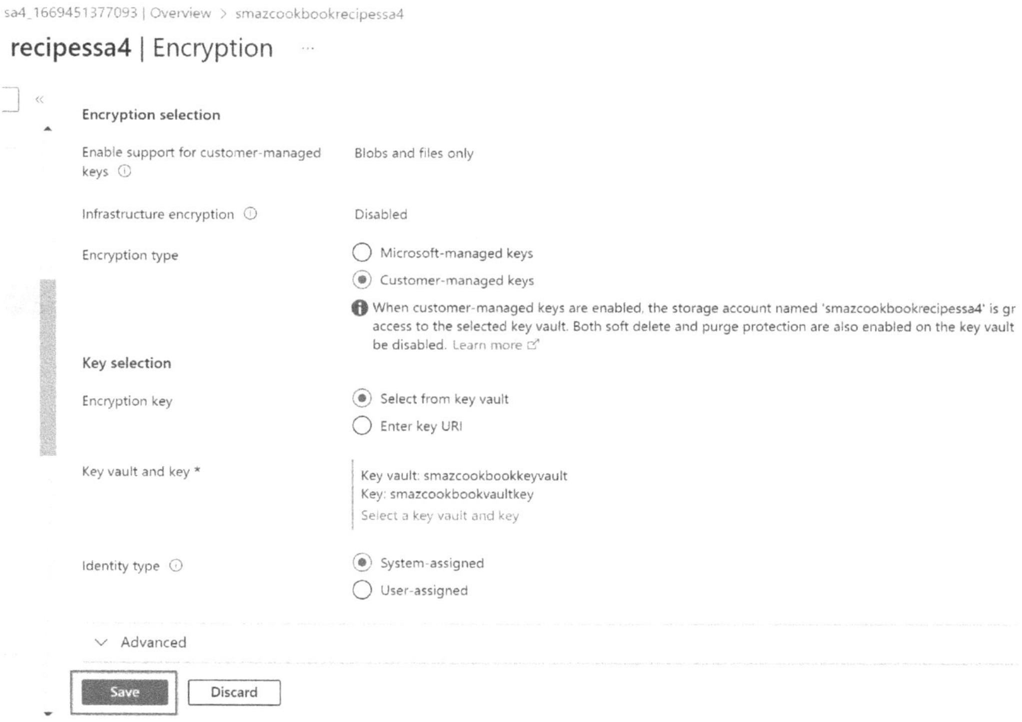

sa4_1669451377093 | Overview > smazcookbookrecipessa4

recipessa4 | Encryption   ⋯

«

Encryption selection

Enable support for customer-managed   Blobs and files only
keys ⓘ

Infrastructure encryption ⓘ   Disabled

Encryption type   ○ Microsoft-managed keys
                                       ◉ Customer-managed keys
                                       ⓘ When customer-managed keys are enabled. the storage account named 'smazcookbookrecipessa4' is gr
                                         access to the selected key vault. Both soft delete and purge protection are also enabled on the key vault
                                         be disabled. Learn more ☐

Key selection

Encryption key   ◉ Select from key vault
                                       ○ Enter key URI

Key vault and key *   Key vault: smazcookbookkeyvault
                                       Key: smazcookbookvaultkey
                                       Select a key vault and key

Identity type ⓘ   ◉ System-assigned
                                       ○ User-assigned

∨ Advanced

[Save]   [Discard]

Figure 6.35 – Save the settings

14. You will receive a notification that the encryption of the storage account was successful.

Figure 6.36 – Encryption successful

15. You can now close the **Encryption** page.

16. Click on **Go to resource** or navigate to your *storage account*.

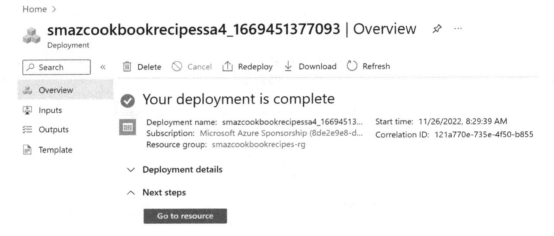

Figure 6.37 – Encryption complete

17. From your *storage account*, click on **Encryption** in the **Security + networking** section and review your customer-managed key information.

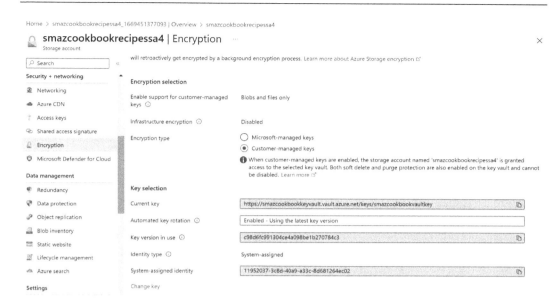

Figure 6.38 – Encryption settings

This task is complete. In the next task, we will clean up the resources created in this recipe.

Task – cleaning up resources

Perform the following steps:

1. In the search box in the *Azure portal*, type `resource groups` and select **Resource Groups** from the listed **Services** results.

2. On the **Resource groups** page, select the *resource group* we created for this recipe, and click **Delete resource group**; this will delete all the resources created as part of this recipe.

Figure 6.39 – Delete resource group

The task of cleaning up the resources created in this recipe is complete.

How it works...

In this final recipe task for this chapter, we looked at implementing encryption for the storage account.

As a *Getting ready* task, we created an Azure storage account for the recipe task to illustrate setting encryption on an existing storage account.

We used Azure Storage Service Encryption using the customer-managed keys encryption option; by default, storage accounts are encrypted by Microsoft-managed keys with no configuration required.

See also

Should you require further information, you can refer to the following Microsoft Learn articles:

- *Configure customer-managed keys in an Azure key vault for an existing storage account*: `https://learn.microsoft.com/en-gb/azure/storage/common/customer-managed-keys-configure-existing-account`

- *What is Azure Key Vault Managed HSM?*: `https://learn.microsoft.com/en-gb/azure/key-vault/managed-hsm/overview`

- *Prevent Shared Key authorization for an Azure Storage account*: `https://learn.microsoft.com/en-gb/azure/storage/common/shared-key-authorization-prevent`

- *Manage storage account access keys*: `https://learn.microsoft.com/en-us/azure/storage/common/storage-account-keys-manage`

- *Assign a Key Vault access policy*: `https://learn.microsoft.com/en-gb/azure/key-vault/general/assign-access-policy`

Part 2: Azure Security Tools

In this part, we will go through recipes that provide complete coverage of the skills and knowledge required to implement and operate native Azure platform security tools.

This part includes the following chapters:

- *Chapter 7, Using Advisor*
- *Chapter 8, Using Microsoft Defender for Cloud*
- *Chapter 9, Using Microsoft Sentinel*
- *Chapter 10, Using Traffic Analytics*

7

Using Advisor

In the previous chapter, we looked at securing the available storage services in Azure. We looked at how the shared responsibility model for public cloud platform providers requires the customer to have implemented the appropriate solutions and controls to ensure their storage is secure and protected.

This previous chapter concluded part one of this cookbook, where we looked at Azure's security features and capabilities as a first-party solution. We covered recipes that provided security for identities, networks, remote access, VMs, databases, and storage.

This first chapter in *Part 2* of the cookbook, which covers Azure Security Tools, looks at the first of these tools, Advisor.

In this chapter, you will learn how to secure and protect Azure environments using the security aspects of the Advisor recommendations engine. Advisor can also provide recommendations for reliability, performance, cost, and operational excellence, although these are beyond the scope of this book.

This section will break down the chapter into sections on security recommendations and secure scores, how to configure security recommendations, configure alerts, and perform remediation of recommendations.

By the end of this chapter, you will have gone through the following recipes and gained the skills to make the most effective use of Advisor:

- Reviewing the security recommendations
- Implementing the security recommendations

Technical requirements

For this chapter, it is already assumed that you have an *Azure AD tenancy* and an *Azure subscription* from completing the recipes in previous chapters of this cookbook. If you skipped straight to this section, the information to create a new *Azure AD tenancy* and an *Azure subscription* for these recipes is included in the following list of requirements.

For this chapter, the following are required:

- A device with a browser, such as *Edge* or *Chrome*, to access the Azure portal at `https://portal.azure.com`.

- An **Azure AD tenancy** and **Azure subscription**; you may use existing ones or sign up for free: `https://azure.microsoft.com/en-us/free`.

- An **Owner role** for the **Azure subscription**.

- The following access permissions: `https://learn.microsoft.com/en-us/azure/advisor/permissions`.

Reviewing the security recommendations

With the eagerness and excitement of moving workloads onto a cloud platform, unfortunately, things can get overlooked, and often governance and control can take a back seat. This can lead to less than optimal deployment, configuration, and operation practices and use of resources, resulting in a security breach and negatively impacting trust in adopting a cloud platform's capabilities. We must not forget the **shared responsibility** we have, along with the cloud platform providers.

This recipe will teach you how to review the security recommendations provided by Advisor to improve your security posture and workload protection.

Getting ready

This recipe requires the following:

- A device with a browser, such as *Edge* or *Chrome*, to access the **Azure portal** at `https://portal.azure.com`

- Access to an **Azure subscription**, where you have access to the **Owner** role for the **Azure subscription**

How to do it...

This task consists of the following tasks:

- Accessing Advisor

- Reviewing Advisor's recommendations

Task – accessing Advisor

Perform the following steps:

1. Sign in to the Azure portal at `https://portal.azure.com`.

2. In the search bar, type `advisor`; click on **Advisor** from the list of services shown:

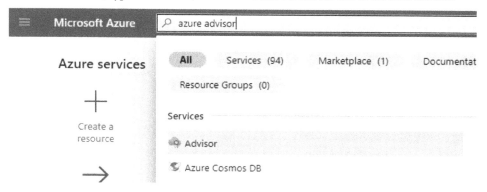

Figure 7.1 – Searching for Advisor

3. When **Advisor** opens, it will load the **Advisor score** page.

This task of accessing Advisor in the Azure portal is complete. In the next task, we will review the security recommendations provided through Advisor.

Task – reviewing Advisor's recommendations

Perform the following steps:

1. From the **Advisor score** page, you will see your *Advisor score*, reflecting Microsoft's best practices. You will also see your **Score by category** information; the **Security** category is of interest for this recipe.

2. There are several Advisor category tiles on the at-a-glance view of your Microsoft best practices posture; our focus is the **Security** tile here:

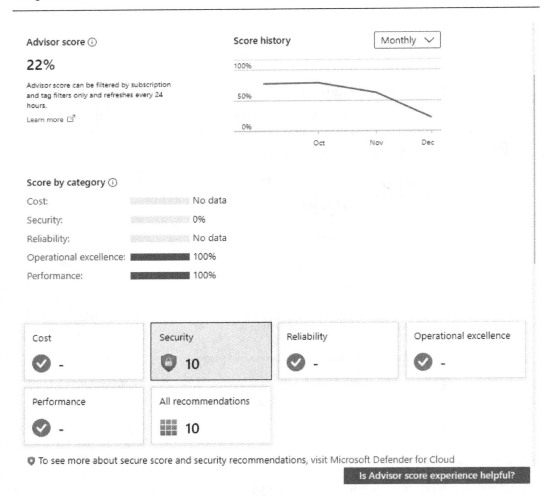

Figure 7.2 – Recommendations dashboard

In the preceding figure, a category score of **100%** indicates that all Advisor-assessed resources follow Advisor's best practices; a category score of **0%** indicates that none of the Advisor-assessed resources follow the Advisor's best practices. The **Security** category tile shows that there are **10** recommendations that we are not following.

3. If you click on the **Security** tile, you will see that it lists these *recommendations*.

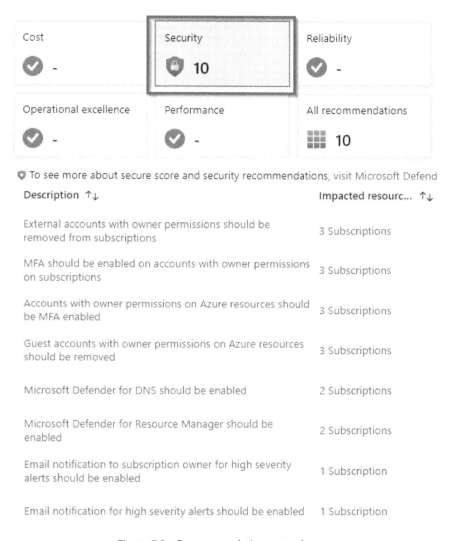

Figure 7.3 – Recommendations at a glance

4.   If we click on the first listed *recommendation*, we will be taken to a blade with further information.

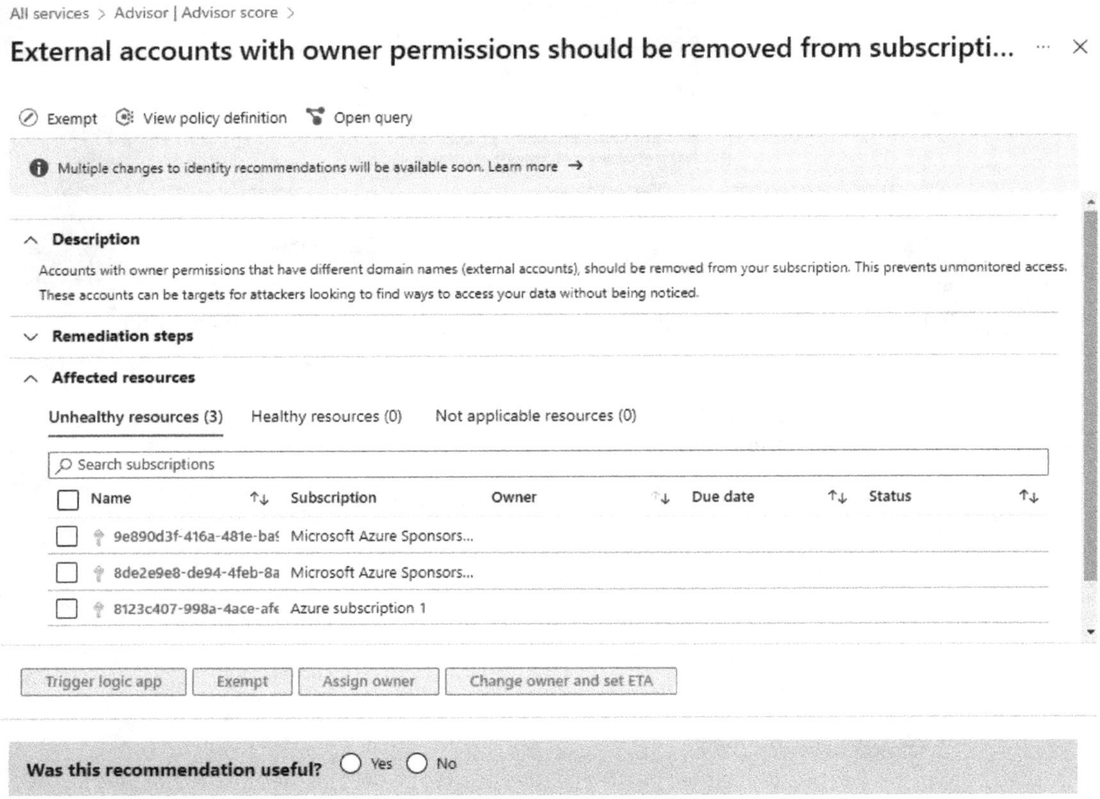

All services > Advisor | Advisor score >

External accounts with owner permissions should be removed from subscripti...   ⋯ ✕

⊘ Exempt   ⊙ View policy definition   🔧 Open query

ℹ️ Multiple changes to identity recommendations will be available soon. Learn more →

∧ **Description**

Accounts with owner permissions that have different domain names (external accounts), should be removed from your subscription. This prevents unmonitored access. These accounts can be targets for attackers looking to find ways to access your data without being noticed.

∨ **Remediation steps**

∧ **Affected resources**

Unhealthy resources (3)   Healthy resources (0)   Not applicable resources (0)

🔍 Search subscriptions

	Name	↑↓	Subscription	Owner	↑↓	Due date	↑↓	Status	↑↓
☐	🔑 9e890d3f-416a-481e-ba9		Microsoft Azure Sponsors...						
☐	🔑 8de2e9e8-de94-4feb-8a		Microsoft Azure Sponsors...						
☐	🔑 8123c407-998a-4ace-afe		Azure subscription 1						

Trigger logic app	Exempt	Assign owner	Change owner and set ETA

Was this recommendation useful?   ◯ Yes ◯ No

Figure 7.4 – Recommendation detail

5. You can also view the **Security** recommendations in the **Recommendations** section in the left-hand menu bar:

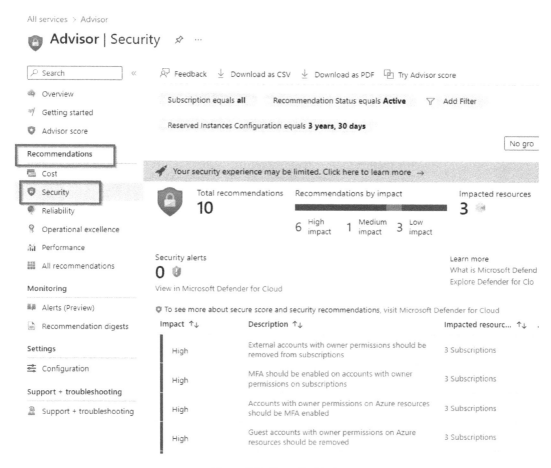

Figure 7.5 – Security blade

6. In the **Security** blade, you can view **Total recommendations**, **Recommendations by impact**, **Impacted resources**, **Security alerts**, and the listed recommendations:

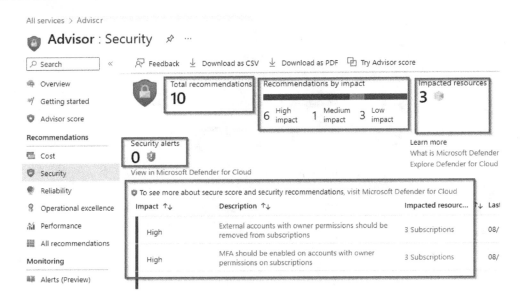

Figure 7.6 – The Security blade

This task of reviewing the security recommendations made available through Advisor is complete.

How it works...

For this recipe, look at how to access the security recommendations available through Advisor to improve your security posture and workload protection.

See also

Should you require further information, you can refer to the following Microsoft Learn articles:

- Advisor Microsoft landing page: `https://azure.microsoft.com/en-us/products/advisor/`

- Advisor documentation: `https://learn.microsoft.com/en-gb/azure/advisor/`

- *Permissions in Azure Advisor*: `https://learn.microsoft.com/en-us/azure/advisor/permissions`

- Microsoft Learn training – *Get started with Advisor*: `https://learn.microsoft.com/en-us/training/modules/get-started-azure-advisor/`

- *Security recommendations - a reference guide*: `https://learn.microsoft.com/en-gb/azure/defender-for-cloud/recommendations-reference`

Implementing the security recommendations

This recipe will teach you how to apply the security recommendations provided by Advisor to improve your security posture and workload protection.

Getting ready

This recipe requires the following:

- A device with a browser, such as *Edge* or *Chrome*, to access the **Azure portal** at https://portal.azure.com

- Access to an **Azure subscription**, where you have access to the **Owner** role for the **Azure subscription**

How to do it...

This task consists of the following tasks:

- Implementing Advisor's recommendations

Task – implementing Advisor's recommendations

Perform the following steps:

1. Sign in to the Azure portal at https://portal.azure.com.

2. In the search bar, type advisor; click on **Advisor** from the list of services shown.

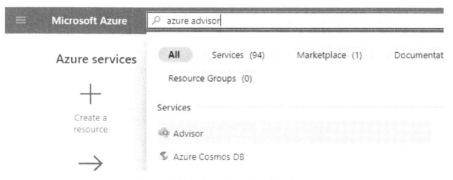

Figure 7.7 – Searching for Advisor

3. When **Advisor** opens, it will load the **Advisor score** page.

4. On the **Advisor score** page, click on the **Security** category tile; this will load the **recommendations** and present them in a list on this page.

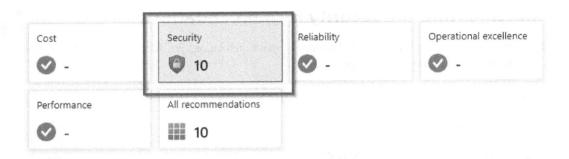

Figure 7.8 – Selecting the Security category tile

5. Select a **recommendation** from the list to remediate; in this scenario, we will select the **Microsoft Defender for Resource Manager should be enabled** recommendation, as represented by the following figure:

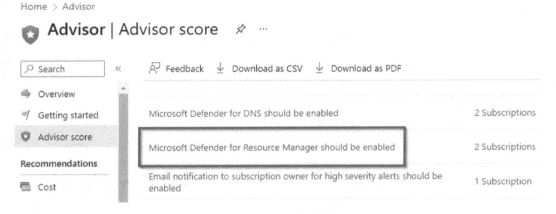

Figure 7.9 – Security recommendation

6. On the **Recommendations** details page, you will see a description of the recommendation, so the recommended best practice. You will also see a **Remediation steps** section; this section can be clicked on and expanded so that you can read the full information on the remediation steps available for this recommendation:

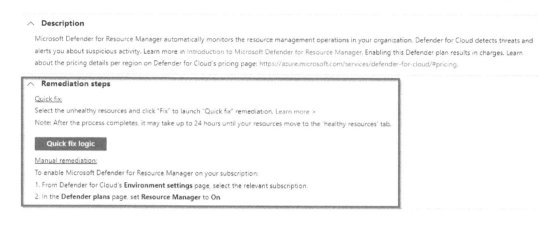

Figure 7.10 – Recommendation remediation steps

7.   If a quick fix is available, then a **Quick fix logic** button will be visible, which you can click to remediate the recommendation. The **Manual remediation** instructions to follow will also be provided.

8.   When you click on the **Quick fix logic** button, an **Automatic remediation script content** blade will pop up, showing the deployment template used to implement the recommendation.

Figure 7.11 – Recommendation script content

9. In the **Affected resources** section, expand and check each **Subscription** item you would like to *apply* this recommendation to, and then click on **Fix**.

Figure 7.12 – Affected resources

10. On the **Fixing resources** pop-up blade, review the information and then click on **Fix 2 resources**, as per our scenario, represented in the following figure:

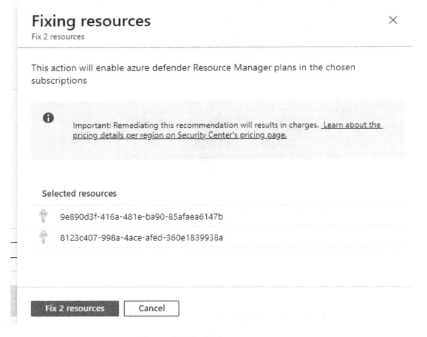

Figure 7.13 – Fixing resources

11. You will be notified that the remediation was successful.

Figure 7.14 – Remediation successful

How it works...

For this recipe, we looked at remediation according to the security recommendations available through Advisor to improve your security posture and workload protection.

See also

- Advisor Microsoft landing page: `https://azure.microsoft.com/en-us/products/advisor/`

- Advisor documentation: `https://learn.microsoft.com/en-gb/azure/advisor/`

- Microsoft Learn training – *Get started with Advisor*: `https://learn.microsoft.com/en-us/training/modules/get-started-azure-advisor/`

8
Using Microsoft Defender for Cloud

In the previous chapter, we covered recipes for using **Azure Advisor** to review security recommendations for your environments and provide alerts and remediation.

This chapter will teach you how to implement **security posture management** and **workload protection** using **Microsoft Defender for Cloud**.

By the end of this chapter, you will have gone through the following recipes to make the most effective use of Microsoft Defender for Cloud:

- Reviewing the components and capabilities of Defender for Cloud
- Enabling enhanced security features of Defender for Cloud
- Adding a Regulatory Standard to the **Regulatory compliance** dashboard
- Assessing your regulatory compliance

Technical requirements

This chapter assumes that you have an **Azure AD tenancy** and an **Azure subscription** from completing the recipes in previous chapters of this cookbook. If you skipped straight to this section, the information needed to create a new Azure AD tenancy and an Azure subscription for these recipes is included in the following list of requirements.

For this chapter, the following is required:

- A device with a browser, such as *Edge* or *Chrome*, to access the **Azure portal:** https://portal.azure.com
- An Azure AD tenancy and Azure subscription; you may use an existing subscription or sign up for free: https://azure.microsoft.com/en-us/free

- An **Owner role** for the Azure subscription

- In addition, you should have the **Security Administrator** role assigned

- **Defender for Cloud** upgraded with **enhanced security features**; these are free for 30 days. More info can be found at the following URL: `https://learn.microsoft.com/en-us/azure/defender-for-cloud/enhanced-security-features-overview`

Terminology reference

We will start with some terminology used with Defender for Cloud:

- **Security operations (SecOps)**: This function deals with managing the inclusion of the day-to-day security monitoring needs of an organization in IT operations.

- **Security posture**: This is the status of an organization's cyber-security measures; its ability to detect and react to security threats.

- **Secure Score** This is a percentage-based score based on Microsoft's best practice security recommendations. It measures your security posture; the higher your score, the greater your security positioning.

- **Cloud security posture management (CSPM)**: This is a means to measure an organization's security posture; a proactive security service that provides a *Secure Score* to measure your security protection levels. It provides actionable recommendations and insights for the remediation of identified threat vectors and vulnerabilities.

- **Cloud workload protection (CWP)**: This refers to the reactive security measure tools that can be implemented to protect workloads identified as potentially at risk from the CSPM insights.

Now that we have covered some related terminology, we will move on to our first recipe for this section.

Review Defender for Cloud components

This recipe will provide you with a high-level overview of the capabilities of Microsoft Defender for Cloud.

Getting ready

This recipe requires the following:

- A device with a browser, such as Edge or Chrome, to access the Azure portal: `https://portal.azure.com`

- Access to an Azure subscription, where you have access to the **Owner** role for the Azure subscription

How to do it...

This task consists of only one step - reviewing the components of Defender for Cloud. Task – review the components of Defender for Cloud

Perform the following steps:

1. Sign in to the Azure portal: `https://portal.azure.com`.

2. In the search bar, type `defender for cloud`; click **Microsoft Defender for Cloud** from the list of services shown.

Figure 8.1 – Search for the Defender for Cloud resource

3. When **Microsoft Defender for Cloud** opens, it will load the **Overview** page.

4. From the top of the dashboard, you will see the following three key metrics presented:

   - **Azure subscriptions**
   - **Assessed resources**
   - **Active recommendations**
   - **Security alerts**

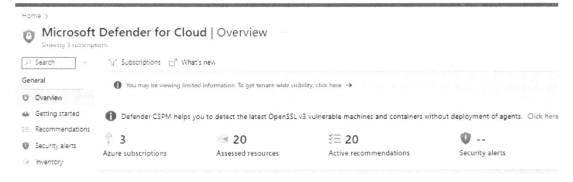

Figure 8.2 – Defender for Cloud overview of key metrics

5. From the **Overview** page for Defender for Cloud, you also will see a dashboard below the key metrics that provides an at-a-glance view of the key components that make up Defender for Cloud; these are as follows:

    *   **Security posture**

        *   **Secure score**

    *   **Regulatory compliance**

    *   **Workload protections**

    *   **Inventory**

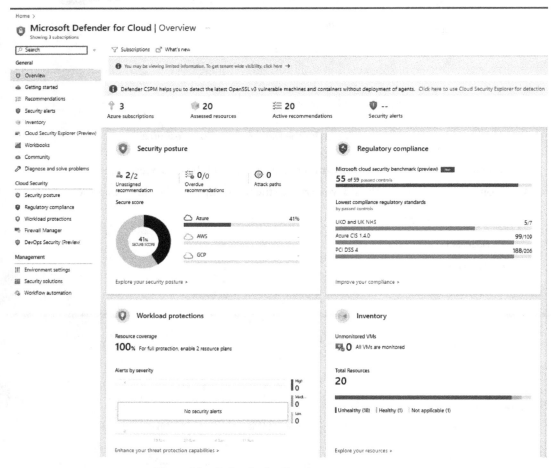

Figure 8.3 – Defender for Cloud core components

This task of reviewing the components of Defender for Cloud is complete.

How it works...

For this recipe, we looked at reviewing the capabilities and components of Defender for Cloud.

See also

Should you wish to learn more about this and related topics, you can refer to the following Microsoft Learn articles:

- Microsoft Defender for Cloud documentation: `https://learn.microsoft.com/en-us/azure/defender-for-cloud/`

- Microsoft Learn training – Introduction to Microsoft Defender for Cloud: `https://learn.microsoft.com/en-us/training/modules/intro-to-defender-cloud/`

- Microsoft Learn training – AZ-500: Manage security operation: `https://learn.microsoft.com/en-us/training/paths/manage-security-operation/`

Enable enhanced security features of Defender for Cloud

This recipe will teach you how to take full advantage of the CWP capabilities provided by enabling the enhanced features of *Microsoft Defender for Cloud*.

The extended *Security Posture* and *Detection* and *Response* features are available to improve your security posture and workload protection. They can be enabled for subscriptions via the paid-for *Defender plans*.

Getting ready

This recipe requires the following to be in place:

- A device with a browser, such as Edge or Chrome, to access the Azure portal: `https://portal.azure.com`.

- Access to an Azure subscription, where you have access to the Owner role for the Azure subscription.

- In addition, you should have the Security Administrator role assigned.

- The subscription should not have the enhanced security features of Microsoft Defender for Cloud already enabled. More info can be found at the following URL: `https://learn.microsoft.com/en-us/azure/defender-for-cloud/enhanced-security-features-overview`.

How to do it...

This task consists of the following steps:

- Enable enhanced security features on a subscription
- Enable enhanced security features on multiple subscriptions

Task – enable enhanced security features on a subscription

Perform the following steps:

1. Sign in to the Azure portal: `https://portal.azure.com`.

2. In the search bar, type `defender for cloud`; click **Microsoft Defender for Cloud** from the list of services shown.

Figure 8.4 – Search for the Defender for Cloud resource

3. When **Microsoft Defender for Cloud** opens, it will load the **Overview** page.

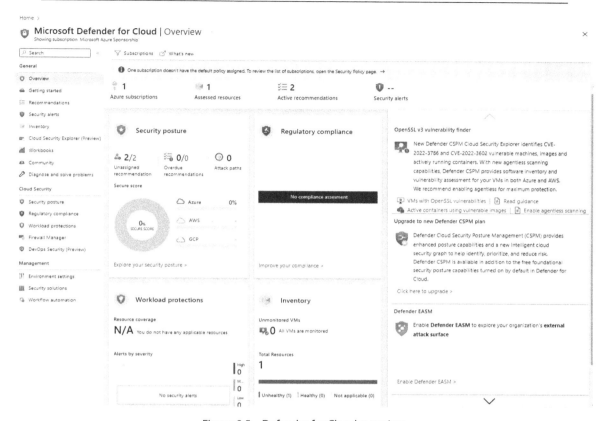

Figure 8.5 – Defender for Cloud overview

4. From the left-hand menu, click **Environment settings** under the **Management** section.

Figure 8.6 – The Environment settings menu option

5. From the **Environment settings** page, click the subscription you wish to enable for *enhanced security features*.

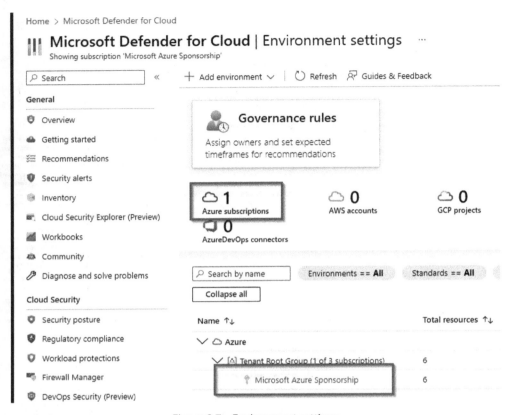

Figure 8.7 – Environment settings

6. From the **Defender plans** page, read the information about the *free CSPM* capabilities available and how you can get the extended features for cloud workloads by enabling the **Defender plans**.

> Defender Cloud Security Posture Management (CSPM) Free - Core posture management capabilities are available for free, covering Multi-Cloud and hybrid environments with continuous assessments, security recommendations, and a unified Secure Score.
>   Enable the below plans to get extended Security Posture and Detection and Response for cloud workloads. Learn more.

Figure 8.8 – Cost information

7. From the bottom of the page, you should also be aware of the pricing implications of enabling Defender plans; you can click **pricing page** to learn more. The direct pricing page URL is `https://azure.microsoft.com/en-us/pricing/details/defender-for-cloud/`.

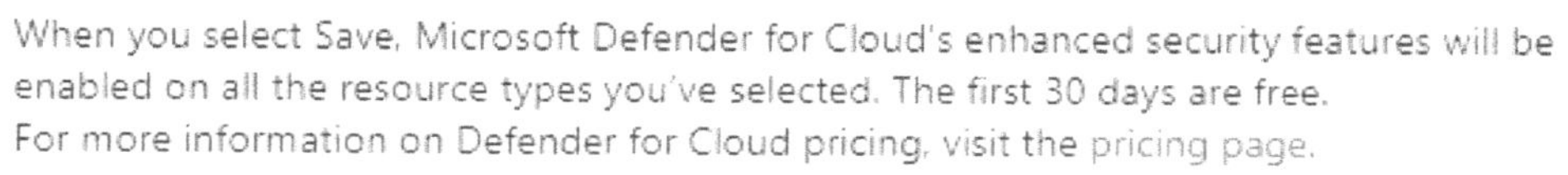

When you select Save, Microsoft Defender for Cloud's enhanced security features will be enabled on all the resource types you've selected. The first 30 days are free.
For more information on Defender for Cloud pricing, visit the pricing page.

Figure 8.9 – Cost information

8.   On the **Defender plans** page, in the **Select Defender Plan** section, click **Enable all**.

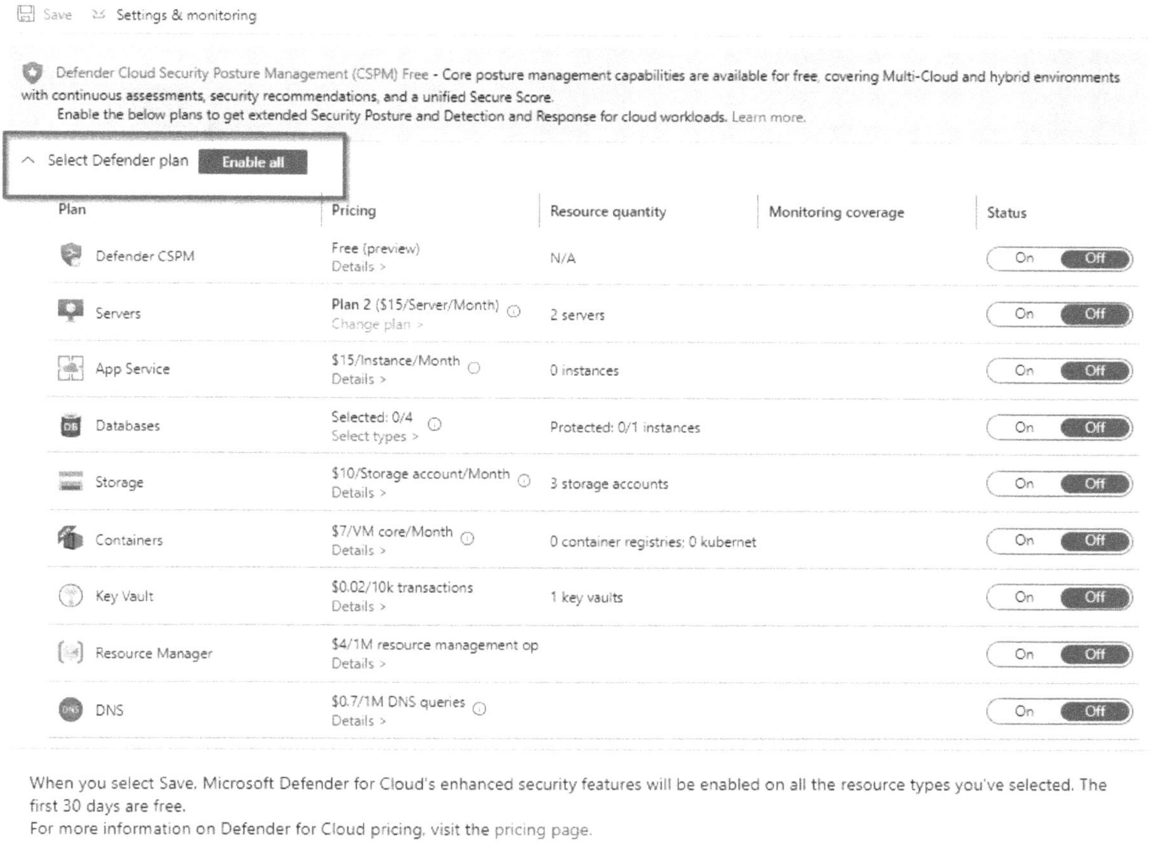

🗔 Save   ↗ Settings & monitoring

🛡 Defender Cloud Security Posture Management (CSPM) Free - Core posture management capabilities are available for free, covering Multi-Cloud and hybrid environments with continuous assessments, security recommendations, and a unified Secure Score.
Enable the below plans to get extended Security Posture and Detection and Response for cloud workloads. Learn more.

∧ Select Defender plan   **Enable all**

Plan	Pricing	Resource quantity	Monitoring coverage	Status
🛡 Defender CSPM	Free (preview) Details >	N/A		On Off
🖥 Servers	Plan 2 ($15/Server/Month) ⓘ Change plan >	2 servers		On Off
🖼 App Service	$15/Instance/Month ⓘ Details >	0 instances		On Off
🗄 Databases	Selected: 0/4 ⓘ Select types >	Protected: 0/1 instances		On Off
🗄 Storage	$10/Storage account/Month ⓘ Details >	3 storage accounts		On Off
📦 Containers	$7/VM core/Month ⓘ Details >	0 container registries: 0 kubernet		On Off
🔑 Key Vault	$0.02/10k transactions Details >	1 key vaults		On Off
📋 Resource Manager	$4/1M resource management op Details >			On Off
🌐 DNS	$0.7/1M DNS queries ⓘ Details >			On Off

When you select Save, Microsoft Defender for Cloud's enhanced security features will be enabled on all the resource types you've selected. The first 30 days are free.
For more information on Defender for Cloud pricing, visit the pricing page.

Figure 8.10 – Defender plans

9.   You will now see that each Plan shows **Monitoring coverage** as **Full** and **Status** as **On**; you may also set each Plan to **On** or **Off** independently as required.

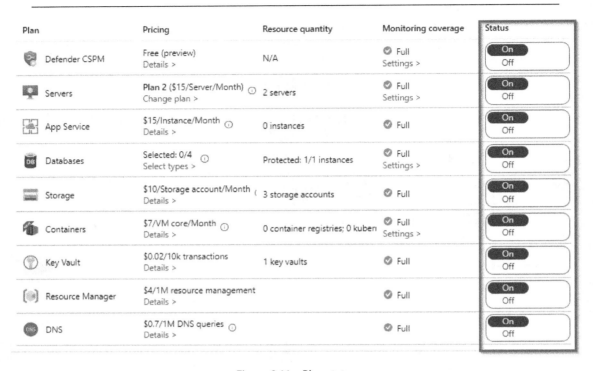

Figure 8.11 – Plan status

10. Click **Save**.

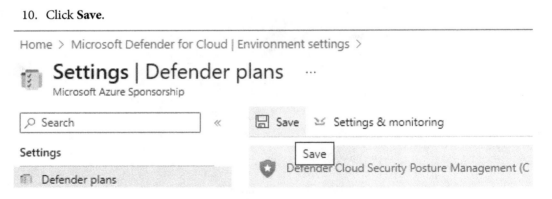

Figure 8.12 – Save settings

11. You will receive a notification that the Defender plans were saved successfully.

Figure 8.13 – Defender for Cloud notification

12. You may now close the **Defender plans** page.

13. Any of the plans can be disabled as required in the future by returning to the **Environment settings** page, selecting your subscription, and from the **Defender plans** page, setting the required Plan to **Off**. Data from the plan will stop being collected; however, the extension won't be uninstalled.

This task to enable enhanced security features of Defender for Cloud on a subscription is complete.

Task – enable enhanced security features on multiple subscriptions

Perform the following steps:

1. Sign in to the Azure portal: `https://portal.azure.com`.

2. In the search bar, type `defender for cloud`; click **Microsoft Defender for Cloud** from the list of services shown.

Figure 8.14 – Search for the Defender for Cloud resource

3. When **Microsoft Defender for Cloud** opens, it will load the **Overview** page.

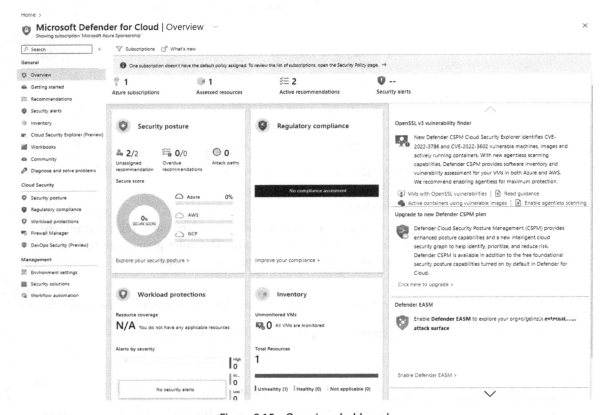

Figure 8.15 – Overview dashboard

4. From the left-hand menu, click **Getting started** under the **General** section.

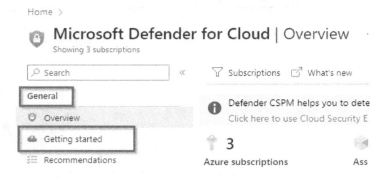

Figure 8.16 – The Getting started menu option

5.   On the **Getting started** page, from the **Upgrade** tab, check the boxes for the subscriptions you would like upgraded, and then click **Upgrade**.

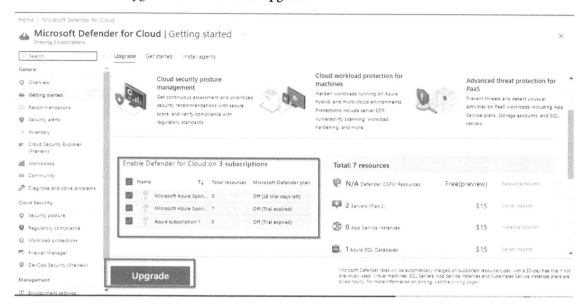

Figure 8.17 – The Getting started page

6.   Note the information regarding the free trial and pricing implications of upgrading; you should click **pricing page** to learn more. The direct pricing page URL is https://azure.microsoft.com/en-us/pricing/details/defender-for-cloud/.

 Microsoft Defender rates will be automatically charged on supported resource types, with a 30-day free trial if not previously used. Virtual machines, SQL Servers, App Service instances and Kubernetes Service instances plans are billed hourly. For more information on pricing, visit the pricing page>

Figure 8.18 – Upgrade costs information

7.   Click **Install agents**.

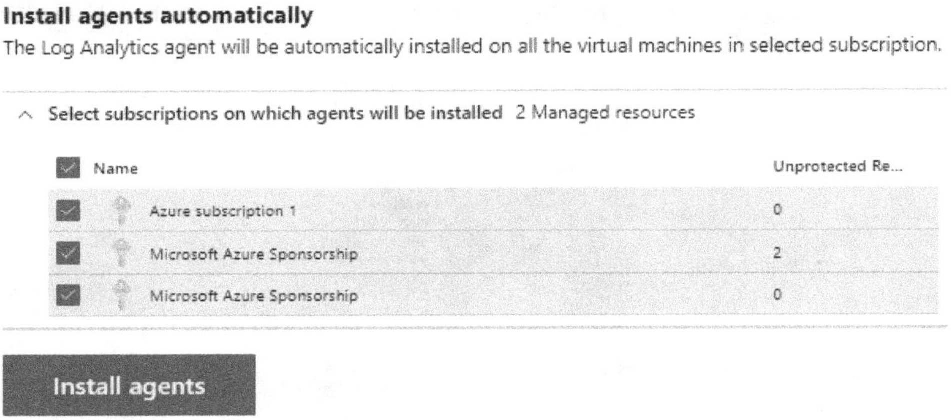

Install agents automatically

The Log Analytics agent will be automatically installed on all the virtual machines in selected subscription.

∧ Select subscriptions on which agents will be installed   2 Managed resources

☑	Name	Unprotected Re...
☑	Azure subscription 1	0
☑	Microsoft Azure Sponsorship	2
☑	Microsoft Azure Sponsorship	0

Install agents

Figure 8.19 – Install agents

8. You will receive a notification that the agents were successfully installed.

Figure 8.20 – Successful installation

9. You may now return to the Defender for Cloud **Overview** page; then click **Environment settings** under the **Management** section, and you will see the plans that have been enabled on each subscription under the **Defender coverage** column.

Figure 8.21 – Enabled plans

10. Any of the plans can be disabled as required in the future by returning to the **Environment settings** page, selecting your subscription, and from the **Defender plans** page, setting the required Plan to **Off**. Data from the plan will stop being collected; however, the extension won't be uninstalled.

This task to enable enhanced security features on multiple subscriptions is complete.

How it works...

Microsoft Defender for Cloud is automatically enabled at no cost for all subscriptions; however, only basic CSPM functionality is provided to ensure a security posture management capability is available as default. You will need to enable the Defender plans, which are paid for, to access the enhanced CWP security features.

For this recipe, we looked at how to take full advantage of the CWP capabilities provided by enabling the enhanced features of Microsoft Defender for Cloud to improve your security posture and workload protection.

See also

Should you wish to learn more about this and related topics, you can refer to the following Microsoft Learn articles:

- Microsoft Defender for Cloud documentation: `https://learn.microsoft.com/en-us/azure/defender-for-cloud/`

- Microsoft Learn training – Introduction to Microsoft Defender for Cloud: `https://learn.microsoft.com/en-us/training/modules/intro-to-defender-cloud/`

- Microsoft Learn training – AZ-500: Manage security operation: `https://learn.microsoft.com/en-us/training/paths/manage-security-operation/`

Add a standard to the Regulatory compliance dashboard

This recipe will teach you how to enable enhanced features of Microsoft Defender for Cloud to improve your security posture and workload protection.

Getting ready

This recipe requires the following to be in place:

- A device with a browser, such as Edge or Chrome, to access the Azure portal: `https://portal.azure.com`.

- Access to an Azure subscription, where you have access to the Owner role for the Azure subscription.

- To access the compliance dashboard and managing standards, you must have **Resource Policy Contributor** and **Security Admin** as minimum roles.

- You must have at least **Reader** (or **Global Reader**) access to view the compliance data; **Security Reader** access will not suffice.

- The subscription should have the enhanced security features of Microsoft Defender for Cloud already enabled.

How to do it...

This task consists of the following step:

- Adding a regulatory compliance standard

Task – adding a regulatory compliance standard

Perform the following steps:

1. Sign in to the Azure portal: `https://portal.azure.com`.

2. Navigate to Defender for Cloud, or in the search bar, type `defender for cloud`; click **Microsoft Defender for Cloud** from the list of services shown.

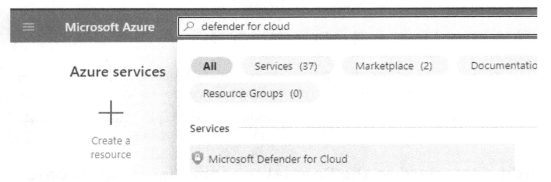

Figure 8.22 – Search for the Defender for Cloud resource

3. When **Microsoft Defender for Cloud** opens, it will load the **Overview** page.

4. From the left-hand menu, click **Regulatory compliance** under the **Cloud Security** section.

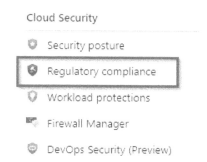

Figure 8.23 – The Regulatory compliance menu option

5.   From the top of the **Regulatory compliance** page, click **Manage compliance policies**.

Figure 8.24 – The Regulatory compliance menu option

6.   From the **Environment settings** page, select the subscription for which to manage the standards compliance posture.

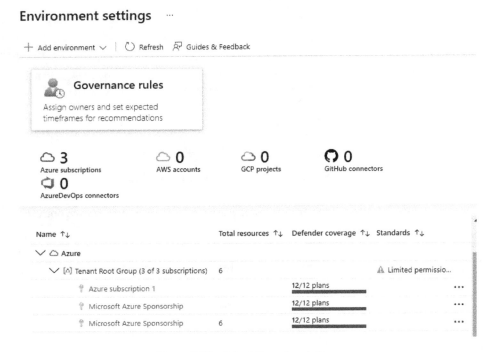

Figure 8.25 – Select the subscription

7. From the **Defender plans** screen, click **Security policy** under the **Policy settings** section.

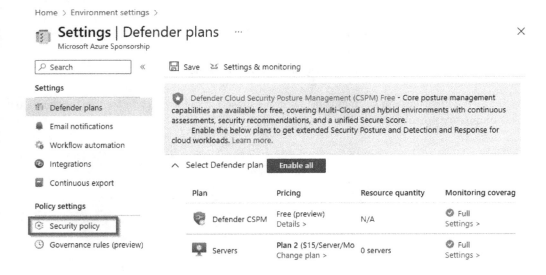

Figure 8.26 – The Security policy menu option

8. From the **Security policy** page, expand the **Industry & regulatory standards** section.

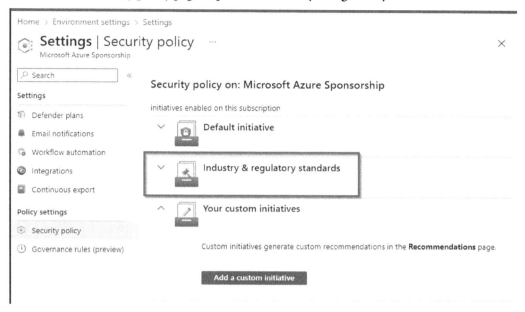

Figure 8.27 – Expand section

9. Review the existing standards, then click **Add more standards**.

Figure 8.28 – Add more standards

10. From the **Add regulatory compliance standards** page, search or locate the required standards to add; click **Add** against the standard.

Figure 8.29 – Add standard

11. From the **Assign Initiative** blade wizard, complete all necessary information for all tabs for that standard, then click **Review + create**.

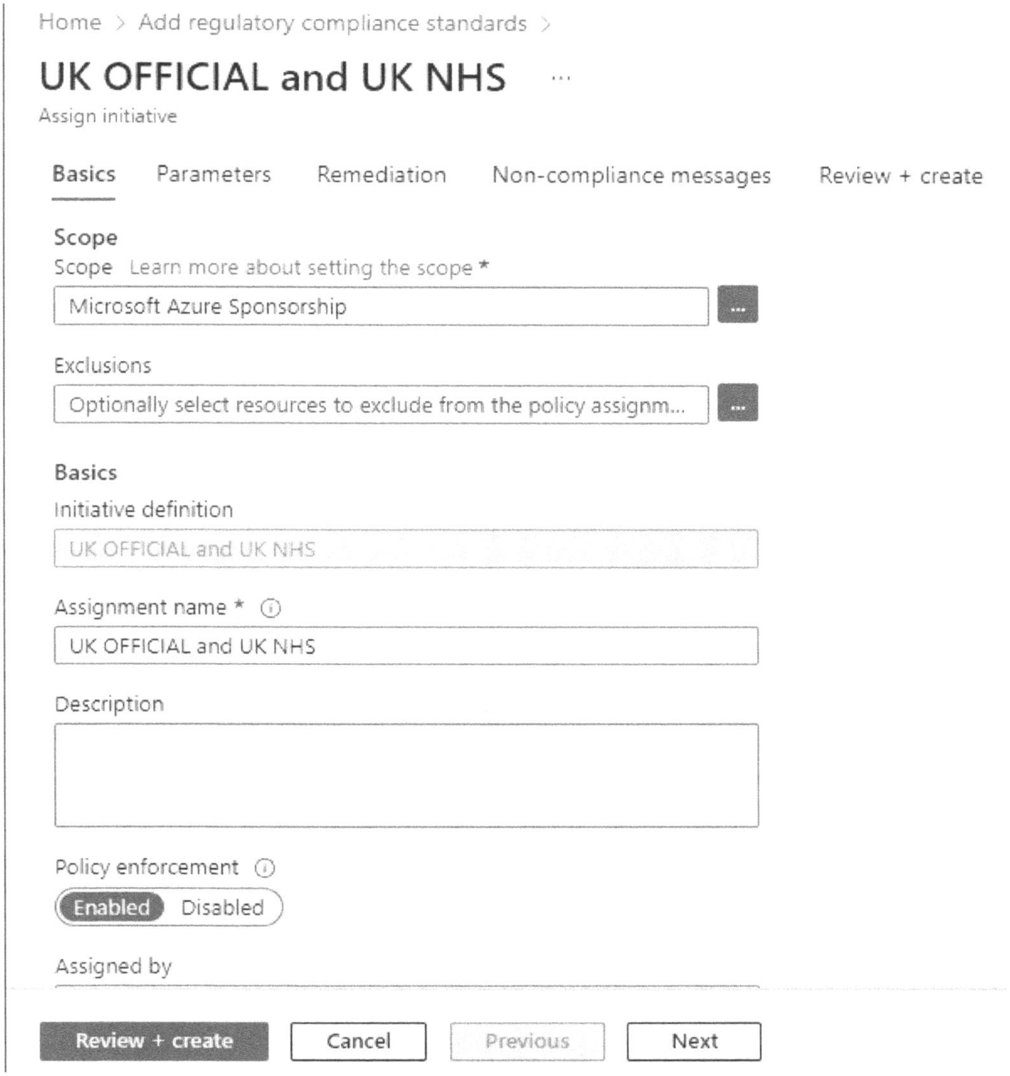

Home > Add regulatory compliance standards >

UK OFFICIAL and UK NHS   ...

Assign initiative

Basics   Parameters   Remediation   Non-compliance messages   Review + create

Scope

Scope   Learn more about setting the scope *

Microsoft Azure Sponsorship   ...

Exclusions

Optionally select resources to exclude from the policy assignm...   ...

Basics

Initiative definition

UK OFFICIAL and UK NHS

Assignment name *   ⓘ

UK OFFICIAL and UK NHS

Description

Policy enforcement   ⓘ

(Enabled   Disabled)

Assigned by

| Review + create | Cancel | Previous | Next |

Figure 8.30 – Review information

12. From the **Review + create** tab, review the information and then click **Create**.

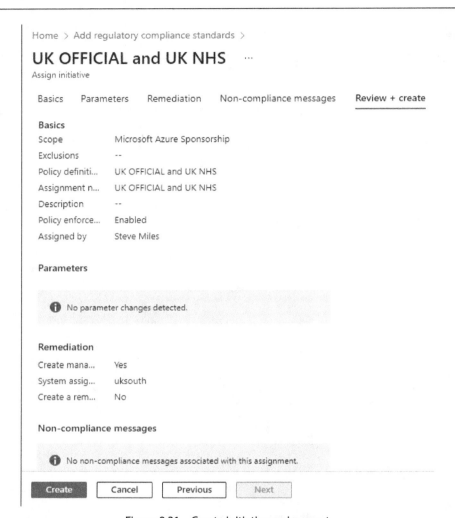

Figure 8.31 – Create initiative assignment

13. You will receive a notification that the creation was successful.

Figure 8.32 – Initiative assignment successful

14. Return to the **Microsoft Defender for Cloud | Regulatory compliance** dashboard page, and after some time, you will see the added standards.

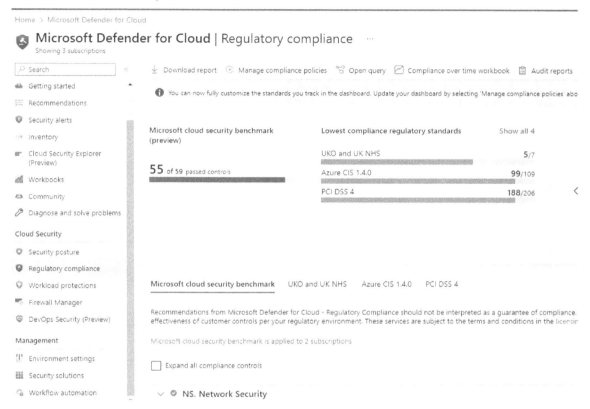

Figure 8.33 – Added standards

15. To *remove* a standard from the **Regulatory compliance** dashboard, navigate to the **Security policy** page and click **Disable** or **Delete** on the standard you wish to remove.

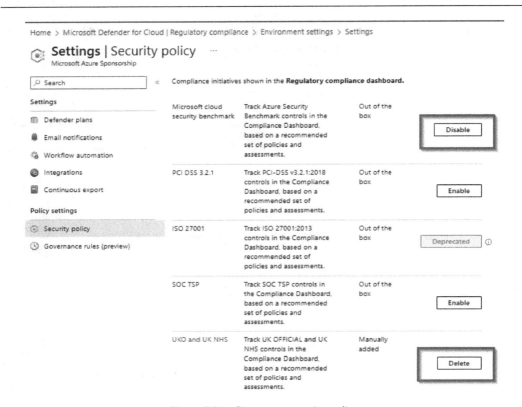

Figure 8.34 – Remove a security policy

This task to add a standard to the **Regulatory compliance** dashboard is complete.

How it works...

The **Regulatory compliance** dashboard in Defender for Cloud, by default, shows the **Microsoft cloud security benchmark** (MCSB). More information on this security benchmark can be found at this URL: https://learn.microsoft.com/en-us/security/benchmark/azure/introduction.

To assess your environment against other standards, they must be added explicitly.

For this recipe, we learned how to add a regulatory compliance standard to Microsoft Defender for Cloud to improve your security posture and workload protection.

To access the **Regulatory compliance** dashboard and manage standards, you must have the **Resource Policy Contributor** and **Security Admin** roles as a minimum.

You must have at least **Reader** (or **Global Reader**) access to view the compliance data; **Security Reader** access will not suffice.

Should you wish to learn more about this and related topics, you can refer to the following Microsoft Learn articles:

- Microsoft Defender for Cloud documentation: `https://learn.microsoft.com/en-us/azure/defender-for-cloud/`

- Microsoft Learn training – Introduction to Microsoft Defender for Cloud: `https://learn.microsoft.com/en-us/training/modules/intro-to-defender-cloud/`

Assess your regulatory compliance

This recipe will teach you how to assess your **regulatory compliance** against a **standard** we added to the previous recipe. Regulatory compliance standards are an enhanced feature of Microsoft Defender for Cloud to improve your security posture and workload protection.

Getting ready

This recipe requires the following to be in place:

- A device with a browser, such as Edge or Chrome, to access the Azure portal: `https://portal.azure.com`

- Access to an Azure subscription, where you have access to the Owner role

- In addition, you should have the **Security Administrator** role assigned

- The subscription should have the *enhanced security features* of Microsoft Defender for Cloud already enabled. More info can be found at the following URL: `https://learn.microsoft.com/en-us/azure/defender-for-cloud/enhanced-security-features-overview`

- The preceding recipe should have been completed, to add a regulatory compliance standard, so your environment can be assessed

- You may wish to create some resources to be assessed, such as a VM, storage account, and so on

How to do it...

This task consists of the following step:

- Assessing a regulatory compliance standard

Task – *assessing a regulatory compliance standard*

Perform the following steps:

1. Sign in to the Azure portal: `https://portal.azure.com`.

2. Navigate to Defender for Cloud, or in the search bar, type `defender for cloud`; click **Microsoft Defender for Cloud** from the list of services shown.

Figure 8.35 – Search for the Defender for Cloud resource

3. When **Microsoft Defender for Cloud** opens, it will load the **Overview** page.

4. From the left-hand menu, click **Regulatory compliance** under the **Cloud Security** section.

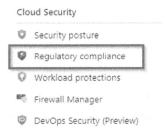

Figure 8.36 – The Regulatory compliance menu option

5. From the **Lowest compliance regulatory standards** section of **Regulatory compliance**, click **Show all 4**. You will see the **Compliance status** section for all the standards that have been added from the previous recipe.

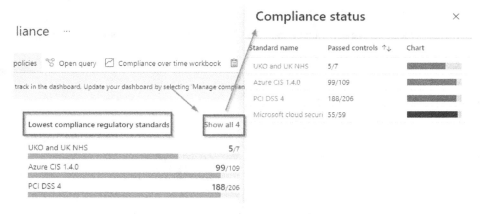

Figure 8.37 – The standard's compliance status

6. To access the standard and review the controls, you can either click on the **Standard name** hyperlink from the **Compliance status** screen or the standard's tab on the **Regulatory compliance** page.

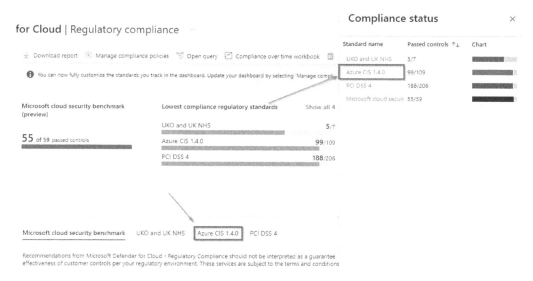

Figure 8.38 – Access the standard you wish to assess

7. From the Standard to assess tab, we can see each compliance control that has been met and the ones that have not.

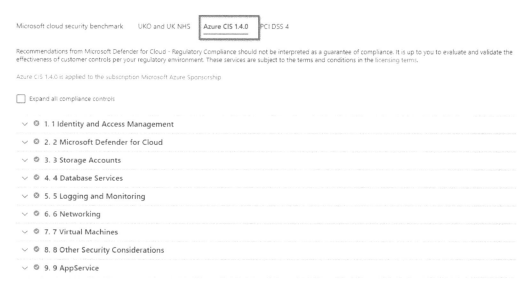

Figure 8.39 – Compliance controls

8. Expand one of the controls that has not been met.

Figure 8.40 – Unmet compliance control

9. The failed resources for that control will be shown. In this example, **multi-factor authentication (MFA)** has not been enabled on a subscription account with **owner permissions**; click on the failed resource **Automated assessments** hyperlink to take you to the details page of this standard's assessment.

Figure 8.41 – Failed control resource

10. Expand the **Remediation steps** section from the **Automated assessments** page of the failed resource and review the information provided; you can use the **Quick Fix** option if this is provided as a remediation step.

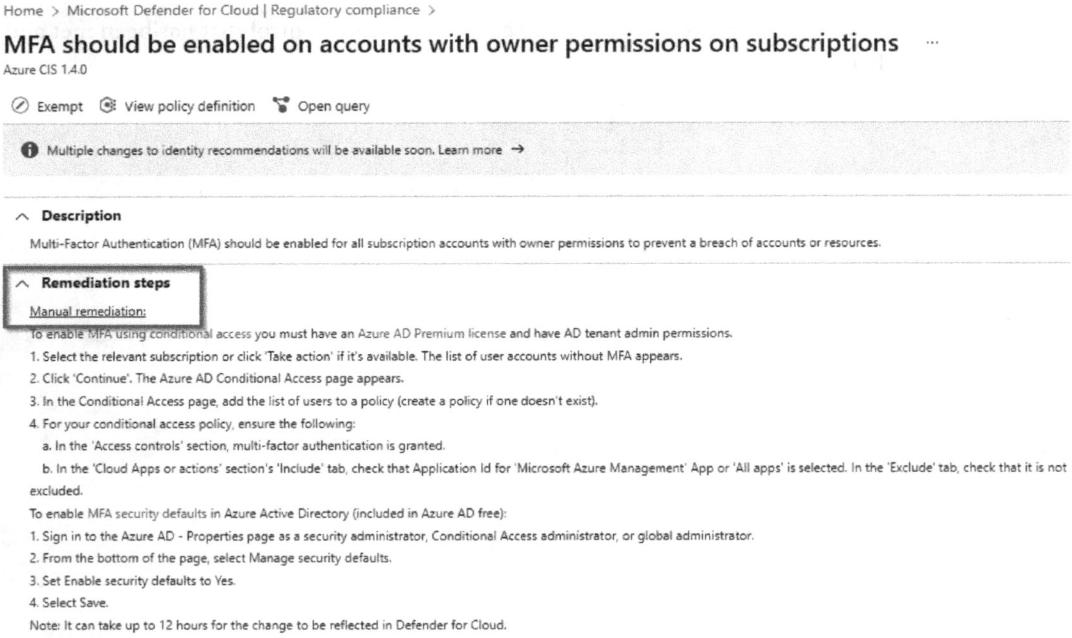

Home > Microsoft Defender for Cloud | Regulatory compliance >

MFA should be enabled on accounts with owner permissions on subscriptions ...
Azure CIS 1.4.0

⊘ Exempt   ⊞ View policy definition   ⌇ Open query

ⓘ Multiple changes to identity recommendations will be available soon. Learn more →

∧ **Description**

Multi-Factor Authentication (MFA) should be enabled for all subscription accounts with owner permissions to prevent a breach of accounts or resources.

∧ **Remediation steps**
Manual remediation:
To enable MFA using conditional access you must have an Azure AD Premium license and have AD tenant admin permissions.
1. Select the relevant subscription or click 'Take action' if it's available. The list of user accounts without MFA appears.
2. Click 'Continue'. The Azure AD Conditional Access page appears.
3. In the Conditional Access page, add the list of users to a policy (create a policy if one doesn't exist).
4. For your conditional access policy, ensure the following:
    a. In the 'Access controls' section, multi-factor authentication is granted.
    b. In the 'Cloud Apps or actions' section's 'Include' tab, check that Application Id for 'Microsoft Azure Management' App or 'All apps' is selected. In the 'Exclude' tab, check that it is not excluded.
To enable MFA security defaults in Azure Active Directory (included in Azure AD free):
1. Sign in to the Azure AD - Properties page as a security administrator, Conditional Access administrator, or global administrator.
2. From the bottom of the page, select Manage security defaults.
3. Set Enable security defaults to Yes.
4. Select Save.
Note: It can take up to 12 hours for the change to be reflected in Defender for Cloud.

Figure 8.42 – Remediation steps

11. You can also expand the **Affected resources** section by clicking the hyperlink of any resources listed under the **Unhealthy resources** section. This will then show **User accounts requiring MFA**; clicking **Continue** allows you to complete the steps outlined in the **Remediation steps** section.

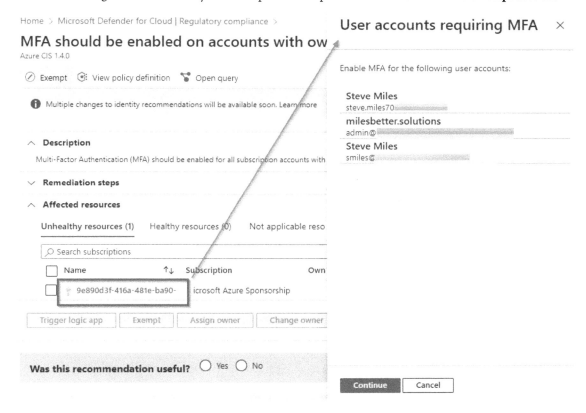

Figure 8.43 – Affected resources

12. You should return to the **Regulatory compliance** dashboard and review each control added to the compliance controls flagged as not met and having failed/unhealthy resources.

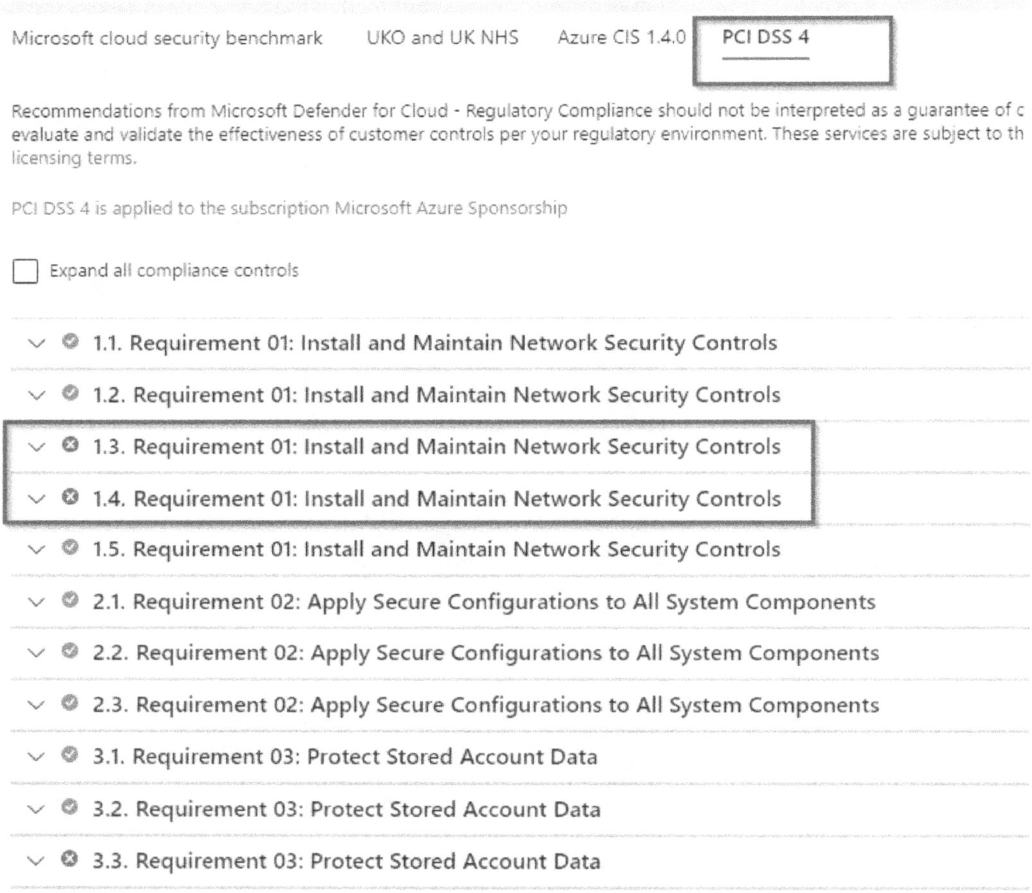

Microsoft cloud security benchmark   UKO and UK NHS   Azure CIS 1.4.0   PCI DSS 4

Recommendations from Microsoft Defender for Cloud - Regulatory Compliance should not be interpreted as a guarantee of c evaluate and validate the effectiveness of customer controls per your regulatory environment. These services are subject to th licensing terms.

PCI DSS 4 is applied to the subscription Microsoft Azure Sponsorship

☐ Expand all compliance controls

∨ ⊘ 1.1. Requirement 01: Install and Maintain Network Security Controls

∨ ⊘ 1.2. Requirement 01: Install and Maintain Network Security Controls

∨ ⊗ 1.3. Requirement 01: Install and Maintain Network Security Controls

∨ ⊗ 1.4. Requirement 01: Install and Maintain Network Security Controls

∨ ⊘ 1.5. Requirement 01: Install and Maintain Network Security Controls

∨ ⊘ 2.1. Requirement 02: Apply Secure Configurations to All System Components

∨ ⊘ 2.2. Requirement 02: Apply Secure Configurations to All System Components

∨ ⊘ 2.3. Requirement 02: Apply Secure Configurations to All System Components

∨ ⊘ 3.1. Requirement 03: Protect Stored Account Data

∨ ⊘ 3.2. Requirement 03: Protect Stored Account Data

∨ ⊗ 3.3. Requirement 03: Protect Stored Account Data

Figure 8.44 – Standards added with failed resource compliance controls

This task to assess regulatory standards that have been added to the **Regulatory compliance** dashboard is complete.

How it works...

For this recipe, we looked at how to assess your regulatory compliance against a standard that we added in the previous recipe.

Regulatory compliance standards are an enhanced feature of Microsoft Defender for Cloud to improve your security posture and workload protection.

You must have at least **Reader** (or **Global Reader**) access to view the compliance data; **Security Reader** access will not suffice.

See also

Should you wish to learn more about this and related topics, you can refer to the following Microsoft Learn articles:

- *Microsoft Defender for Cloud documentation*: `https://learn.microsoft.com/en-us/azure/defender-for-cloud/`

- *Tutorial: Improve your regulatory compliance*: `https://learn.microsoft.com/en-us/azure/defender-for-cloud/regulatory-compliance-dashboard`

- Microsoft Learn training – *Introduction to Microsoft Defender for Cloud*: `https://learn.microsoft.com/en-us/training/modules/intro-to-defender-cloud/`

Using Microsoft Sentinel

The previous chapter covered recipes for enabling and using **Microsoft Defender for Cloud**.

This chapter will teach you how to implement **Microsoft Sentinel**, collect data, set up security alerts through analytics, and create automated responses.

By the end of this chapter, you will have learned the following recipes to make the most effective use of Microsoft Sentinel:

- Reviewing the components of Microsoft Sentinel
- Enabling Microsoft Sentinel
- Creating automation
- Setting up a data connector and analytics rule

Technical requirements

This chapter assumes that you have an **Azure AD tenancy** and an **Azure subscription** from completing the recipes in previous chapters of this cookbook. If you skipped straight to this section, the information needed to create a new Azure AD tenancy and an Azure subscription for these recipes is included in the following list of requirements.

For this chapter, the following are required:

- A device with a browser, such as *Edge* or *Chrome*, to access the Azure portal `https://portal.azure.com`.
- An Azure AD tenancy and Azure subscription; you may use existing or sign up for free: `https://azure.microsoft.com/en-us/free`.
- An **Owner role** for the Azure subscription.
- A **Global Administrator** or **Security Administrator** role on the **Tenant Azure AD**.

Terminology reference

We will start with the terminology used with Microsoft Sentinel:

- **Security operations (SecOps)**: This function deals with managing the incorporation of day-to-day security monitoring needs of an organization into IT operations.

- **Security information and event management (SIEM)**: A SIEM system analyzes real-time security data and events. It can be used as a single pane of glass and a birds-eye view of security operations across an organization's estate; it is often utilized as the primary tool to support a **security operations center (SoC)** and **Soc-as-a-Service**.

- **Security orchestration, automation, and response (SOAR)**: SOAR works with SIEM to provide a full solution set of capabilities that allow the automation and orchestration of responses once critical incidents occur.

Now that we have covered the terminology, we will move on to our first recipe for this section.

Enabling Microsoft Sentinel

Sentinel is Microsoft's cloud-based SIEM and SOAR tool; it is a complete solution that can provide security and event data aggregation, threat analysis, and response across public cloud, hybrid, and on-premises environments.

This recipe will teach you to enable Microsoft Sentinel in your environment.

Getting ready

This recipe requires the following:

- A device with a browser, such as Edge or Chrome, to access the Azure portal: `https://portal.azure.com`

- Access to an Azure subscription, where you have access to the Owner role

How to do it...

This task consists of the following step:

- Enabling Microsoft Sentinel

Task – enabling Microsoft Sentinel

Perform the following steps:

1. Sign in to the Azure portal: `https://portal.azure.com`.

2. In the search bar, type `Microsoft Sentinel`; select **Microsoft Sentinel** from the list of services shown.

Figure 9.1 – Search for the Microsoft Sentinel resource

3. When **Microsoft Sentinel** opens, click on **Create** from the top-menu bar.

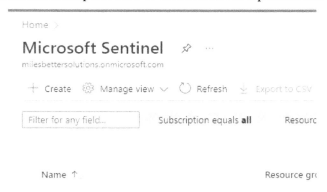

Figure 9.2 – Create a Sentinel instance

4. From the **Add Microsoft Sentinel to a workspace** page, review the information regarding the **31-day free trial**, then click **Create a new workspace**.

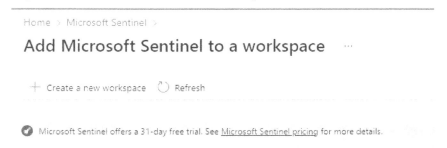

Figure 9.3 – Create a new workspace for Microsoft Sentinel

5. From the **Basics** tab of the **Create Log Analytics workspace** page, fill in the required information for **Project Details** and **Instance details** and click **Review + Create**.

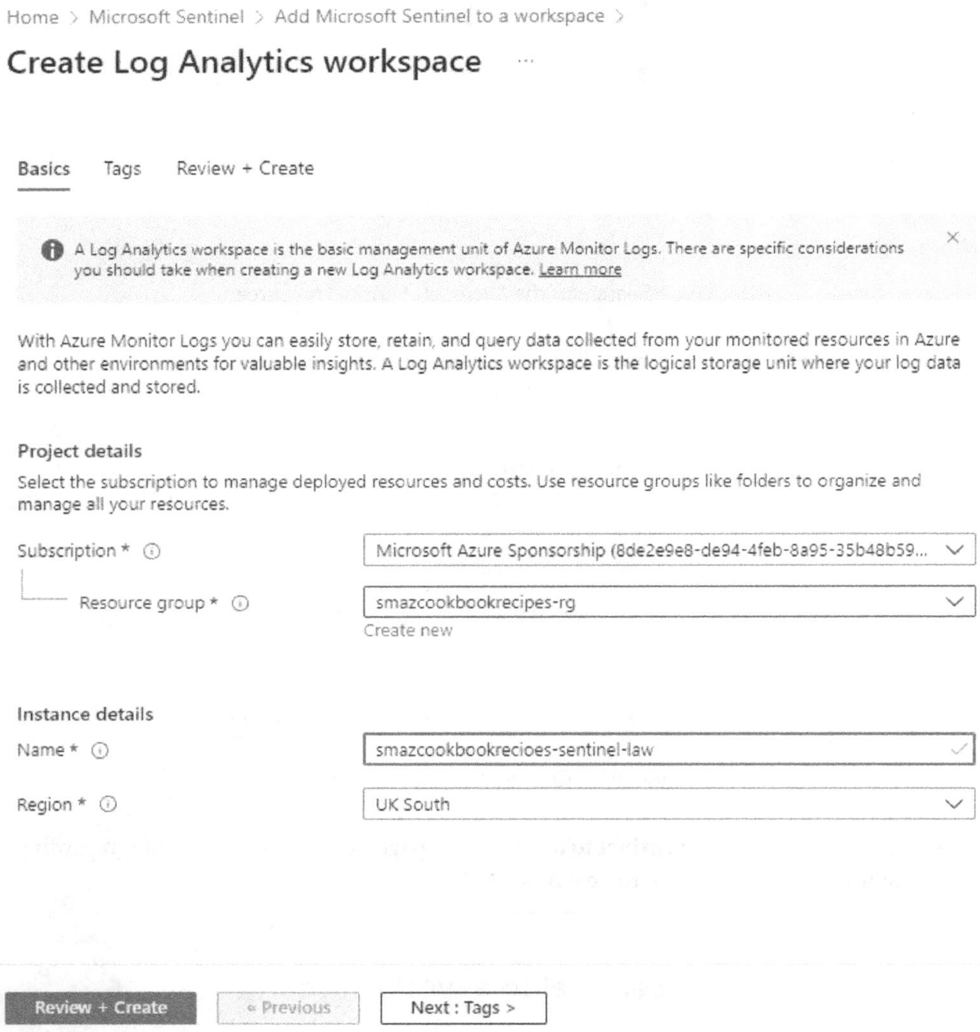

Home > Microsoft Sentinel > Add Microsoft Sentinel to a workspace >

Create Log Analytics workspace   ...

Basics   Tags   Review + Create

> ℹ A Log Analytics workspace is the basic management unit of Azure Monitor Logs. There are specific considerations
> you should take when creating a new Log Analytics workspace. Learn more   ✕

With Azure Monitor Logs you can easily store, retain, and query data collected from your monitored resources in Azure and other environments for valuable insights. A Log Analytics workspace is the logical storage unit where your log data is collected and stored.

Project details

Select the subscription to manage deployed resources and costs. Use resource groups like folders to organize and manage all your resources.

Subscription * ⓘ | Microsoft Azure Sponsorship (8de2e9e8-de94-4feb-8a95-35b48b59... ∨ |

└─── Resource group * ⓘ | smazcookbookrecipes-rg   ∨ |
Create new

Instance details

Name * ⓘ | smazcookbookrecioes-sentinel-law   ✓ |

Region * ⓘ | UK South   ∨ |

| Review + Create | | « Previous | | Next : Tags > |

Figure 9.4 – Create Log Analytics workspace

6. From the **Review + Create** tab, review the information and click **Create**.

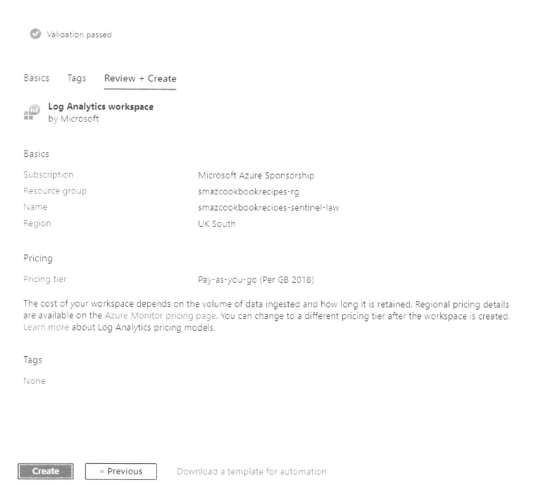

Figure 9.5 – Create workspace validation

7. You will receive a notification that the deployment succeeded; click **Add**.

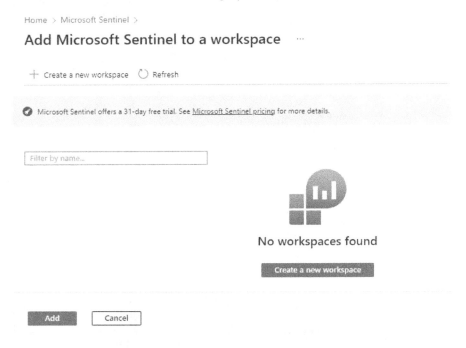

Figure 9.6 – Add Sentinel to a workspace

8. You will receive a **Successfully added Microsoft Sentinel** notification to confirm that it was added to the created **Log Analytics workspace**.

Figure 9.7 – Notification of success

9. You will be redirected to the **Microsoft Sentinel** page and presented with a page banner displaying a **Microsoft Sentinel free trial activated** message. You must review the billing information provided in the **Learn more** hyperlink. Then click **OK** to dismiss the banner.

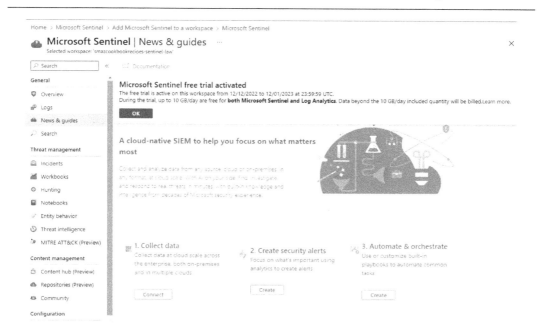

Figure 9.8 – Sentinel landing page

10. You will be directed to the **Get started** tab on the **News & guides** page in the **General** section of Microsoft Sentinel.

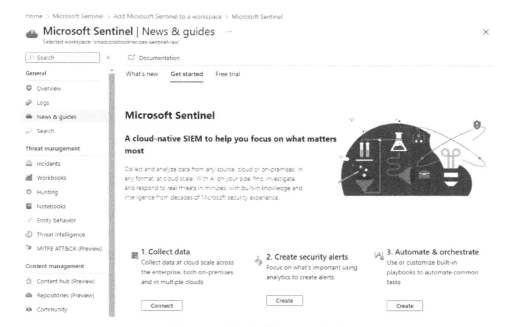

Figure 9.9 – Sentinel Get started tab

11. On the **Get started** tab, you will see three keys steps that we will follow in the remaining recipes in this chapter; for clarity, they are as follows:

- **Collect data**

- **Create security alerts**

- **Automate & orchestrate**

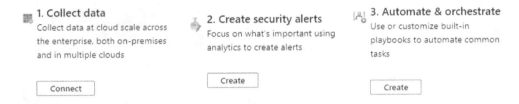

Figure 9.10 – Sentinel Get started steps

This task to enable Microsoft Sentinel is completed.

How it works...

Microsoft Sentinel collects security signaling (security log data) and examines this signal data for patterns that may indicate an attack; this then correlates event information to identify potentially abnormal activity. Issues that are identified create an automated alert response, and remediation is carried out. This relationship is represented in the following diagram:

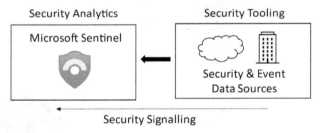

Figure 9.11 – Sentinel positioning

Azure Sentinel provides the following core capabilities:

- **Collects**: Security data is collected across an organization

- **Detects**: Threats are detected through AI-powered threat intelligence

- **Investigates**: Threat-generated critical incidents are investigated

- **Responds**: Responses are generated through automated reactions and remediations

The end-to-end *security operations* capabilities of Sentinel are represented in the following diagram:

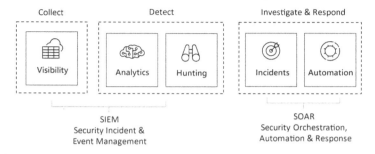

Figure 9.12 – Sentinel security operations capabilities

See also

Should you wish to learn more about this and related topics, you can refer to the following Microsoft Learn articles:

- Microsoft Sentinel documentation: `https://learn.microsoft.com/en-us/azure/sentinel/`

- Quickstart: Onboard Microsoft Sentinel: `https://learn.microsoft.com/en-us/azure/sentinel/quickstart-onboard`

- Microsoft Learn training modules and paths: `https://learn.microsoft.com/en-us/training/browse/?expanded=azure&products=azure-sentinel`

Reviewing Microsoft Sentinel components

Now that we have learned how to enable Microsoft Sentinel in your environment, this recipe will provide you with a high-level overview of its capabilities and components.

As a cloud-based SIEM and SOAR solution, Sentinel can act as the tooling to support a SOC and SOC-as-a-Service approach.

Getting ready

This recipe requires the following:

- A device with a browser, such as Edge or Chrome, to access the Azure portal: `https://portal.azure.com`

- Access to an Azure subscription, where you have access to the Owner role

- The subscription should have Microsoft Sentinel enabled

How to do it...

This task consists of the following step:

- Review the Microsoft Sentinel components

Task – Microsoft Sentinel

Perform the following steps:

1. Sign in to the Azure portal: `https://portal.azure.com`.
2. From the search bar, type `Microsoft Sentinel`; click **Microsoft Sentinel** from the list of services shown.

Figure 9.13 – Search for the Microsoft Sentinel resource

3. When Sentinel opens, click on the workspace created in the previous recipe.

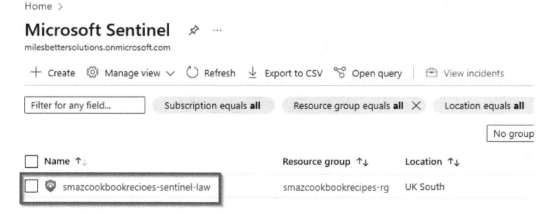

Figure 9.14 – Select the Sentinel workspace

4. When the Sentinel workspace opens, you will see the **Selected workspace** you are viewing. The left-hand menu bar is categorized into the following settings:

    - **General**

    - **Threat management**

    - **Content management**

    - **Configuration**

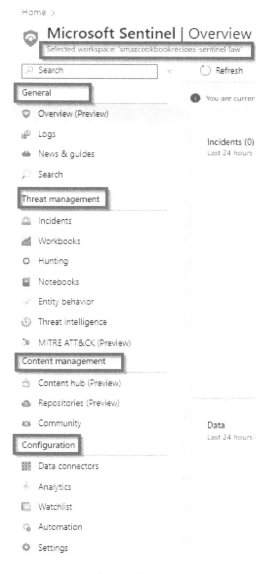

Figure 9.15 – The Workspace menu

5. The **Overview (Preview)** page, under the **General** section, is split into four main operational area tiles as follows:

- **Incidents**

- **Automation**

- **Data**

- **Analytics** (*alert rules*)

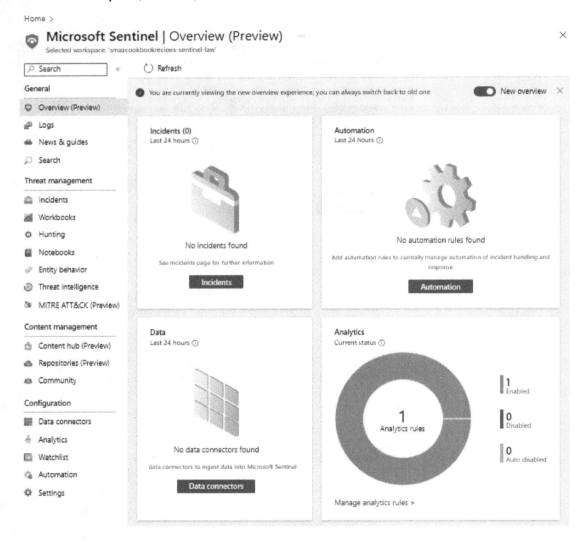

Figure 9.16 – Sentinel operational areas

6. Under **Threat management**, click **Incidents** and review the information presented by the capabilities of this area of Sentinel.

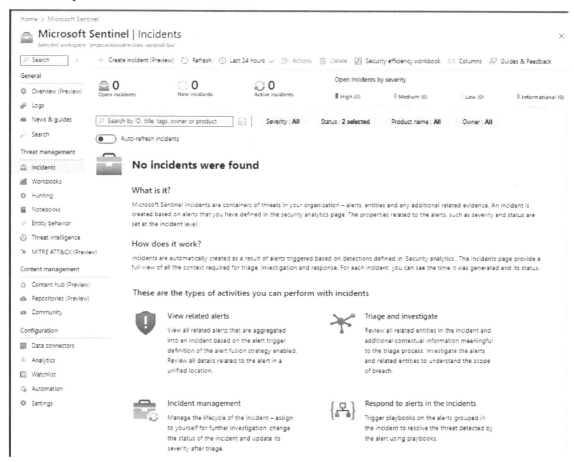

Figure 9.17 – Incidents

7. Under **Configuration**, click **Automation** and review the information presented by the capabilities of this area of Sentinel.

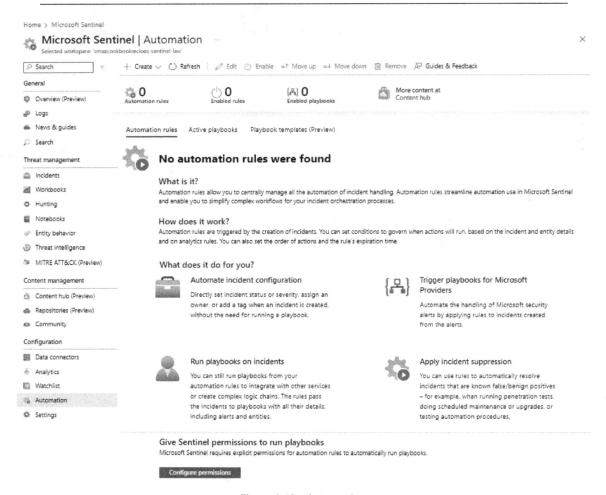

Figure 9.18 – Automation

8. Under **Configuration**, click **Data connectors** and review the information presented by the capabilities of this area of Sentinel.

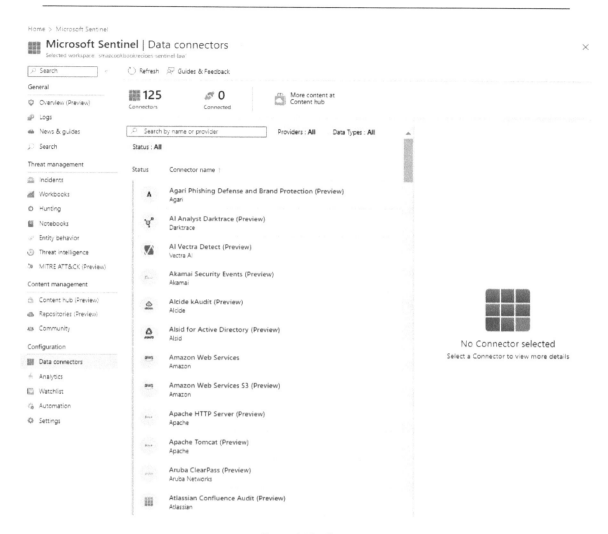

Figure 9.19 – Data

9. Under **Configuration**, click **Analytics** and review the information presented by the capabilities of this area of Sentinel.

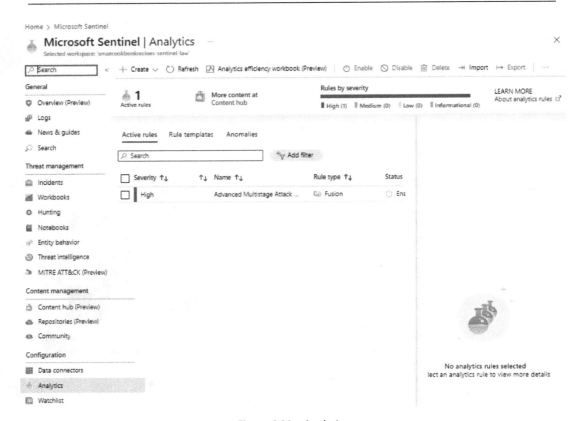

Figure 9.20 – Analytics

This task to review the components and capabilities of Microsoft Sentinel is completed.

How it works...

For this recipe, we reviewed the components and capabilities of Microsoft Sentinel as a cloud-based SIEM and SOAR solution that can act as the tooling to support a SOC and SOC-as-a-Service approach.

See also

Should you wish to learn more about this and related topics, you can refer to the following Microsoft Learn articles:

- Microsoft Sentinel documentation: `https://learn.microsoft.com/en-us/azure/sentinel/`

- Microsoft Learn training modules and paths: `https://learn.microsoft.com/en-us/training/browse/?expanded=azure&products=azure-sentinel`

- What is Microsoft Sentinel?: `https://learn.microsoft.com/en-gb/azure/sentinel/overview`

Creating automation

Now that we have learned how to enable Azure Sentinel in your environment, this recipe will teach you how to set up playbooks so that you can automate responses to incidents that we will trigger from data connector signal data and alert log rules that we will set up in the following recipe in this chapter.

Getting ready

This recipe requires the following to be in place:

- A device with a browser, such as Edge or Chrome, to access the Azure portal: `https://portal.azure.com`

- Access to an Azure subscription, where you have access to the Owner role

- The subscription should have Microsoft Sentinel enabled

How to do it...

This task consists of the following step:

- Creating a playbook

Task – creating a playbook

Perform the following steps:

1. Sign in to the Azure portal: `https://portal.azure.com`.

2. From the search bar, type `Microsoft Sentinel`; click **Microsoft Sentinel** from the list of services shown.

Figure 9.21 – Search for the Sentinel resource

3. When Sentinel opens, click on the created workspace from the previous recipe.

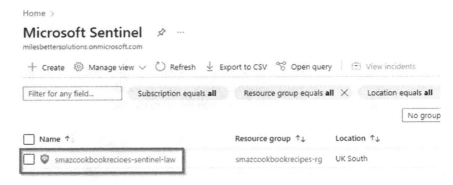

Figure 9.22 – Select the Sentinel workspace

4. From the Sentinel workspace, click **News & guides** under the **General** section; then, from the **Get started** tab, click **Create** from the **Automate & orchestrate** step.

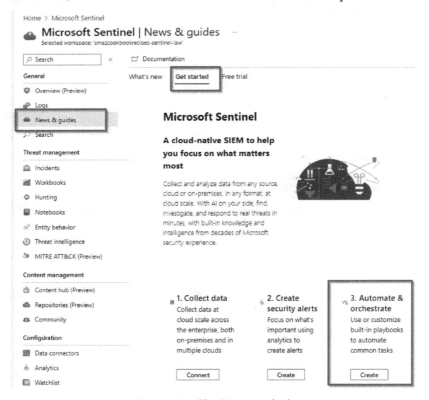

Figure 9.23 – The Get started tab

5. From the **Automation** page, click the **Playbook templates (preview)** tab.

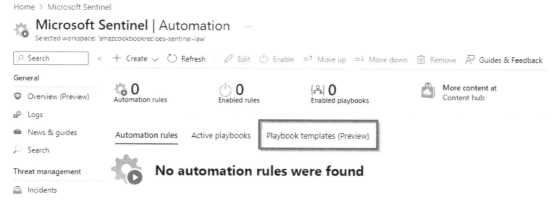

Figure 9.24 – Automation

6. In the list of available playbook templates on the **Playbook templates (preview)** tab, search for and open the **Send email with formatted incident report** template; then click **Create playbook**.

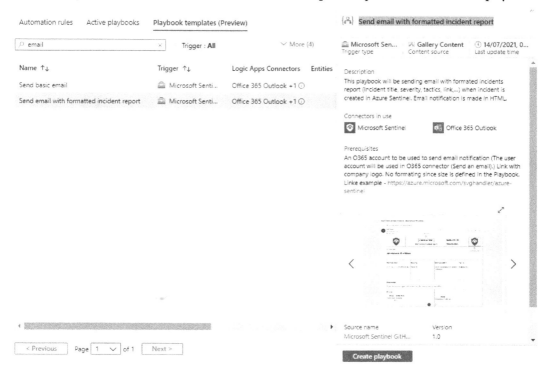

Figure 9.25 – Create a playbook

7. On the **Basics** tab, click **Next : Parameters >**, then from the **Parameters** tab, enter the required information in the **NotificationEmail** field, and then either leave the default or change the **Company logo link** and **Company name** fields. Then, click **Next : Connections >**.

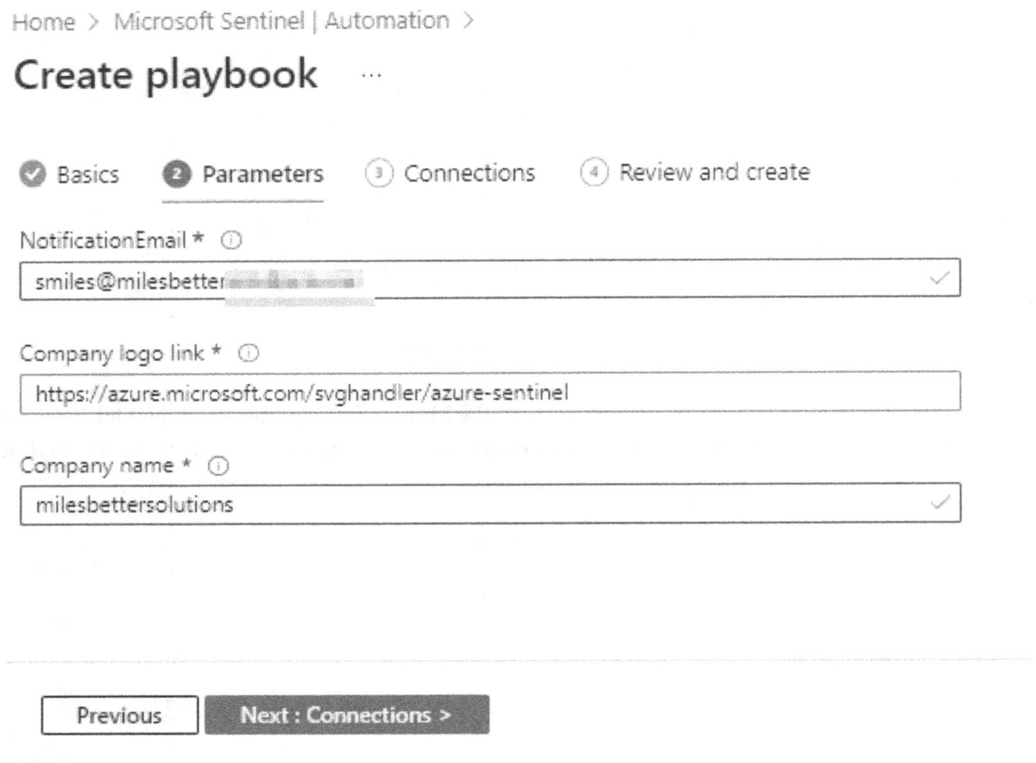

Home > Microsoft Sentinel | Automation >

Create playbook   ...

✓ Basics   ② Parameters   ③ Connections   ④ Review and create

NotificationEmail * ⓘ

smiles@milesbetter▮▮▮▮▮▮▮▮▮   ✓

Company logo link * ⓘ

https://azure.microsoft.com/svghandler/azure-sentinel

Company name * ⓘ

milesbettersolutions   ✓

[Previous]   [Next : Connections >]

Figure 9.26 – The Create playbook parameters tab

8. On the **Connections** tab, click **Next : Review and create >**.

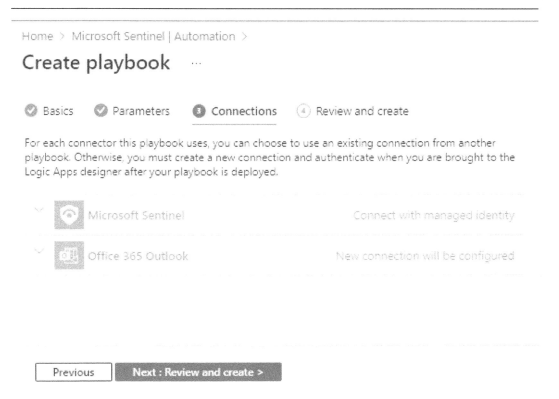

Figure 9.27 – The Create playbook Connections tab

9.   On the **Review and create** tab, click **Create and continue to designer**.

Home > Microsoft Sentinel | Automation >

Create playbook ···

✓ Basics   ✓ Parameters   ✓ Connections   ④ Review and create

Basics

Subscription	Microsoft Azure Sponsorship
Resource group	smazcookbookrecipes-rg
Region	UK South
Playbook name	Send-email-with-formatted-incident-report
Diagnostics logs workspace	Disabled
Integration service environment	Disabled

Parameters

NotificationEmail	▬▬▬▬▬▬▬▬▬
Company logo link	https://azure.microsoft.com/svghandler/azure-sentinel
Company name	MilesBetter SOC

[Previous] [**Create and continue to designer**]

Figure 9.28 – Create a playbook review

10. After creating the playbook, we must complete two further actions: **grant permissions to run the playbook** and **authorize the mail connection**. We will step through these actions in this recipe.

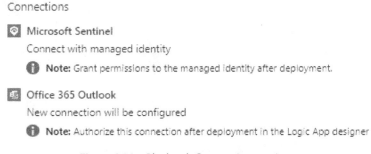

Connections

◎ Microsoft Sentinel
Connect with managed identity
ⓘ **Note:** Grant permissions to the managed identity after deployment.

📧 Office 365 Outlook
New connection will be configured
ⓘ **Note:** Authorize this connection after deployment in the Logic App designer

Figure 9.29 – Playbook Connections actions

11. From the **Logic app designer**, you need to authorize the mail connection and validate the M365 email address selected to receive the email response. Click the information symbol, which will trigger a pop-up window to prompt you to complete signing in with the account used with the *M365 email address.*

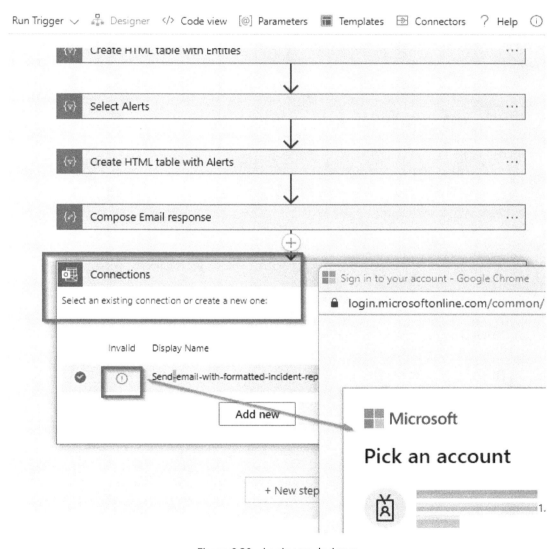

Figure 9.30 – Logic app designer

12. If the *M365 email address* was validated successfully, the **Logic app designer** steps should appear, as represented in the following screenshot:

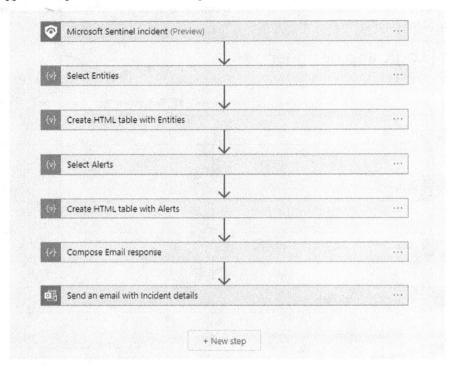

Figure 9.31 – Logic app designer steps

13. You will now see the playbook created on the **Automation** page on the **Active playbooks** tab.

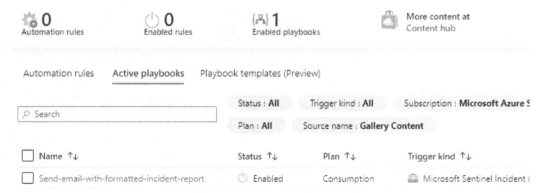

Figure 9.32 – Active playbooks

14. Next, we need to **assign permissions** for Sentinel to the playbook. In Sentinel, click on **Settings** in the **Configuration** section.

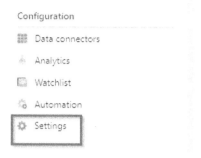

Figure 9.33 – Sentinel settings

15. Click on the **Settings** tab of the **Settings** page.

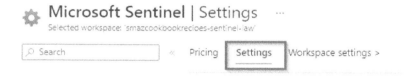

Figure 9.34 – The Settings tab

16. Expand the **Playbook permissions** section on the **Settings** page, and click on **Configure permissions**.

⌃ Playbook permissions

What is it?
Automation rules allow you to centrally manage all the automation of incident handling. Automation rules streamline automation use in Microsoft Sentinel and enable you to simplify complex workflows for your incident orchestration processes.

Playbook permissions
Microsoft Sentinel automation rules can run Logic App playbooks to integrate with other services or create complex logic chains for incident handling. Explicit permissions are required to use this functionality.

Figure 9.35 – Configure permissions

17. Select the resource groups on the **Manage permissions** blade that contains the playbooks that Sentinel can run, then click **Apply**.

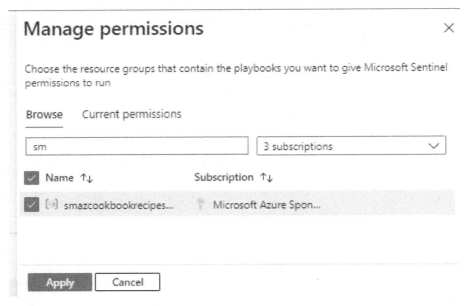

Figure 9.36 – Manage permissions

18. You will receive a notification that the permissions were added.

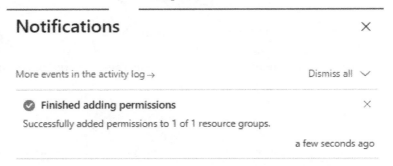

Figure 9.37 – Notification of success

This task to create a playbook is completed.

How it works...

For this recipe, we created a playbook for use with incidents to provide an automated response via email. We will use this playbook for the analytics rule we create in the next recipe.

See also

Should you wish to learn more about this and related topics, you can refer to the following Microsoft Learn articles:

- Microsoft Sentinel documentation: `https://learn.microsoft.com/en-us/azure/sentinel/`

- Microsoft Learn training modules and paths: `https://learn.microsoft.com/en-us/training/browse/?expanded=azure&products=azure-sentinel`

- Tutorial: Use playbooks with automation rules in Microsoft Sentinel: `https://learn.microsoft.com/en-us/azure/sentinel/tutorial-respond-threats-playbook`

Set up data connectors

Now that we have learned how to enable Azure Sentinel in your environment, this recipe will teach you how to set up data connectors so that you can start collecting signal information for analysis and alerting.

Getting ready

This recipe requires the following to be in place:

- A device with a browser, such as *Edge* or *Chrome*, to access the Azure portal: `https://portal.azure.com`

- Access to an Azure subscription, where you have access to the Owner role

- The subscription should have Microsoft Sentinel enabled

How to do it...

This task consists of the following steps:

- Gather signal data using a data connector

- Create an analytics rule

Task – gather signal data using data connectors

Perform the following steps:

1. Sign in to the Azure portal: `https://portal.azure.com`.

2. In the search bar, type `Microsoft Sentinel`; click **Microsoft Sentinel** from the list of services shown.

Figure 9.38 – Search for the Sentinel resource

3. When Sentinel opens, click on the workspace created in the previous recipe.

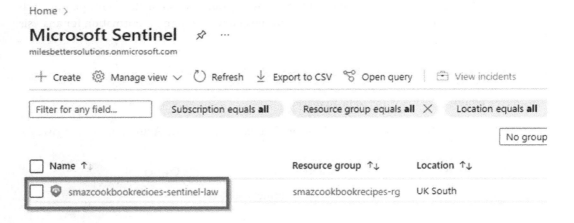

Figure 9.39 – Select workspace

4. In the Sentinel workspace, click **News & guides** in the **General** section; then, from the **Get started** tab, click **Connect** from the **Collect data** step.

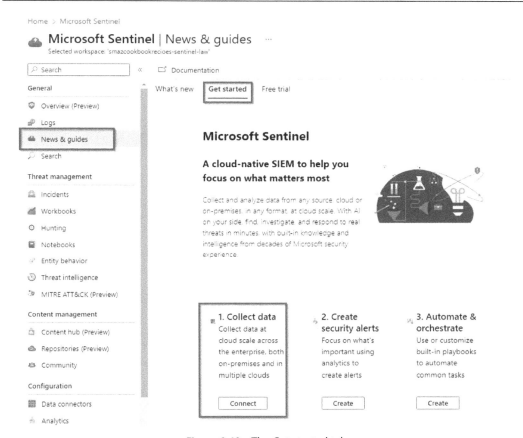

Figure 9.40 – The Get started tab

5.   On the **Data connectors** page, we will search for and then add data connectors.

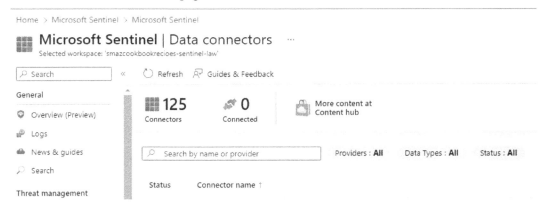

Figure 9.41 – Data connectors

6. In the search box, type `azure active directory` and click the **Azure Active Directory** *data connector* listed in the search results.

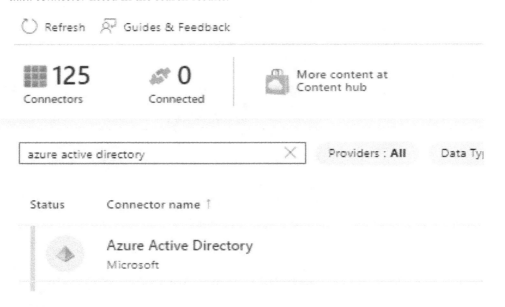

Figure 9.42 – Data connector

7. On the **Connector** blade that opens, click the **Open connector** page.

Figure 9.43 – Data connector information

8. On the **Connector** page, check the boxes for all **log types** to be collected for analysis by Sentinel, then click **Apply Changes**.

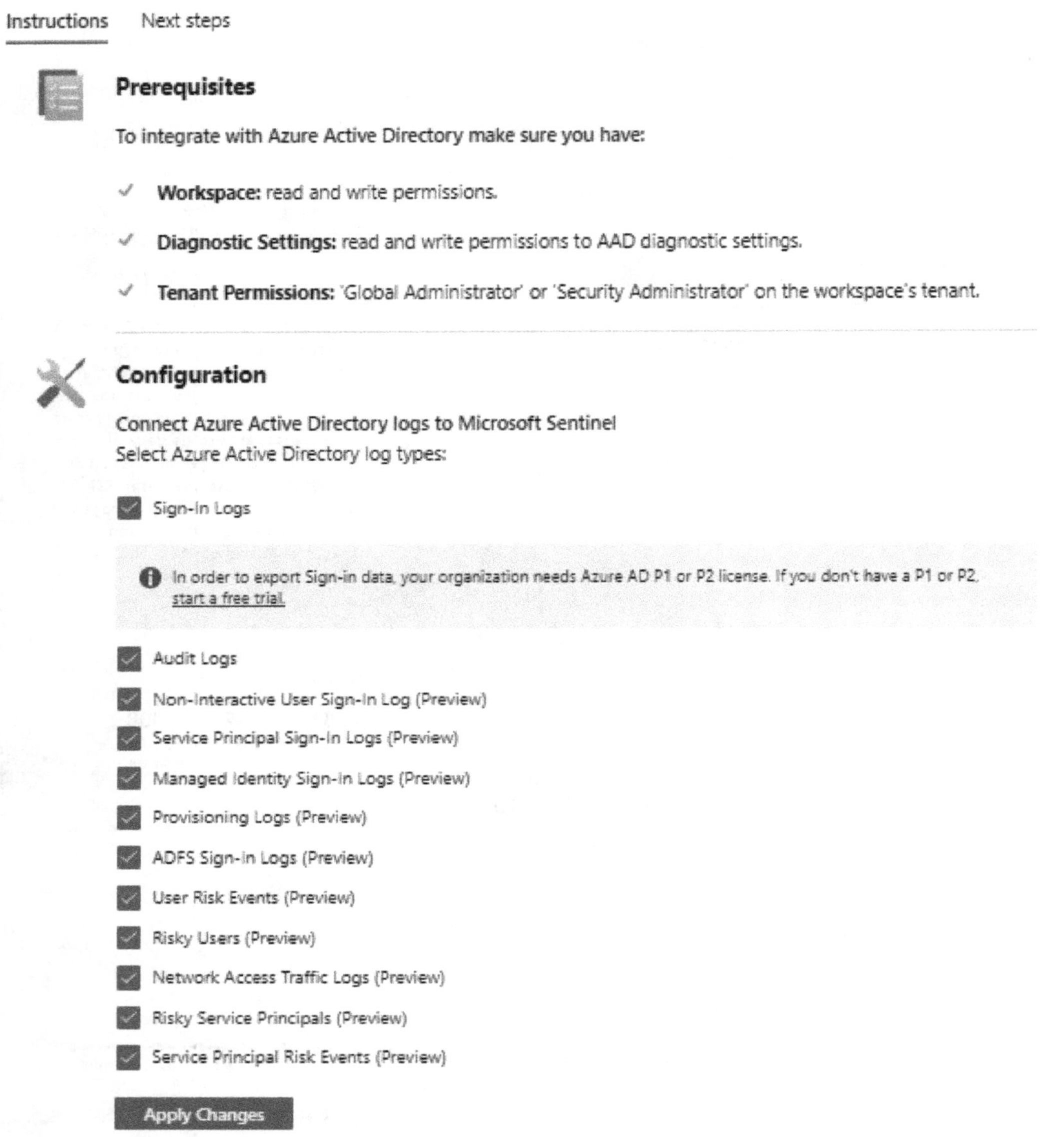

Instructions   Next steps

Prerequisites

To integrate with Azure Active Directory make sure you have:

✓   **Workspace:** read and write permissions.

✓   **Diagnostic Settings:** read and write permissions to AAD diagnostic settings.

✓   **Tenant Permissions:** 'Global Administrator' or 'Security Administrator' on the workspace's tenant.

Configuration

Connect Azure Active Directory logs to Microsoft Sentinel

Select Azure Active Directory log types:

☑ Sign-In Logs

> ⓘ In order to export Sign-in data, your organization needs Azure AD P1 or P2 license. If you don't have a P1 or P2, start a free trial.

☑ Audit Logs

☑ Non-Interactive User Sign-In Log (Preview)

☑ Service Principal Sign-In Logs (Preview)

☑ Managed Identity Sign-In Logs (Preview)

☑ Provisioning Logs (Preview)

☑ ADFS Sign-In Logs (Preview)

☑ User Risk Events (Preview)

☑ Risky Users (Preview)

☑ Network Access Traffic Logs (Preview)

☑ Risky Service Principals (Preview)

☑ Service Principal Risk Events (Preview)

Apply Changes

Figure 9.44 – Data connector configuration

9.   You will receive a notification that the changes were successfully applied.

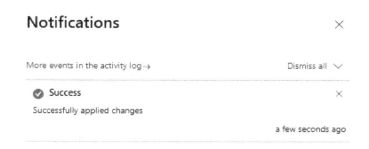

Figure 9.45 – Notification of success

This task to connect a data connector is completed.

Task – create an analytics rule

Perform the following steps:

1. On the **Azure AD Connector** page created in the previous recipe, click the **Next steps** tab.

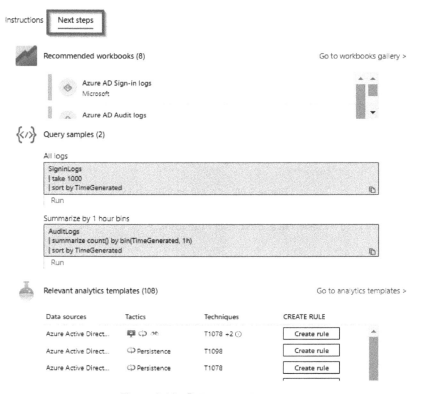

Figure 9.46 – Data connector page

2. On the **Next Steps** tab of the connector, you should review all the activities you would like to be alerted about.

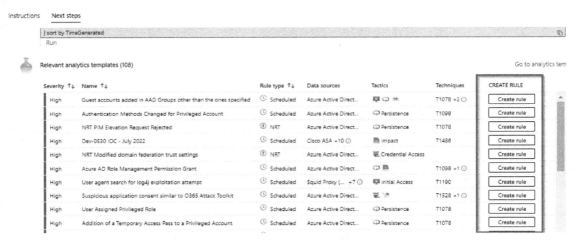

Figure 9.47 – Analytics templates

3. For our example, we wish to use the **Failed login attempts to Azure Portal** rule; to do so, click on **Create rule** against the analytics template's Name value.

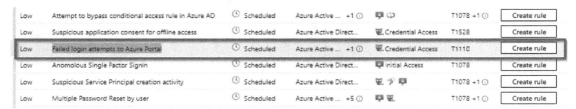

Figure 9.48 – Create analytics rule

4. In **Analytics rule wizard**, click **Next : Set rule logic >**.

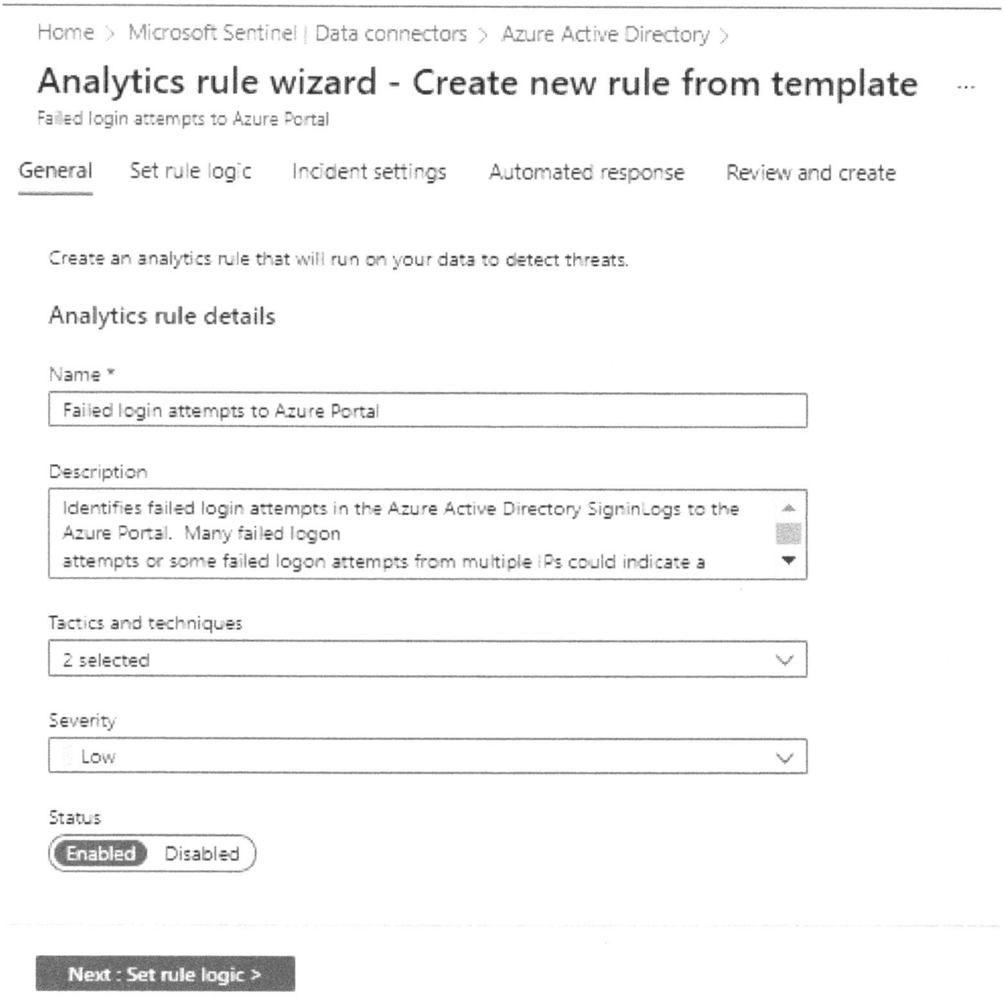

Home > Microsoft Sentinel | Data connectors > Azure Active Directory >

Analytics rule wizard - Create new rule from template ...
Failed login attempts to Azure Portal

General   Set rule logic   Incident settings   Automated response   Review and create

Create an analytics rule that will run on your data to detect threats.

Analytics rule details

Name *

```
Failed login attempts to Azure Portal
```

Description

```
Identifies failed login attempts in the Azure Active Directory SigninLogs to the
Azure Portal.  Many failed logon
attempts or some failed logon attempts from multiple IPs could indicate a
```

Tactics and techniques

```
2 selected                                                              ∨
```

Severity

```
Low                                                                     ∨
```

Status

(**Enabled** Disabled)

Next : Set rule logic >

Figure 9.49 – Analytics rule wizard

5. Review the information on the **Set rule logic** tab in the **Rule query** tab. For an incident to
    trigger, a threshold must be met, which can be edited as required. For this recipe, edit the
    values as follows:

    - `let threshold_Failed = 1;`

    - `let threshold_FailedwithSingleIP = 1;`

    - `let threshold_IPAddressCount = 1;`

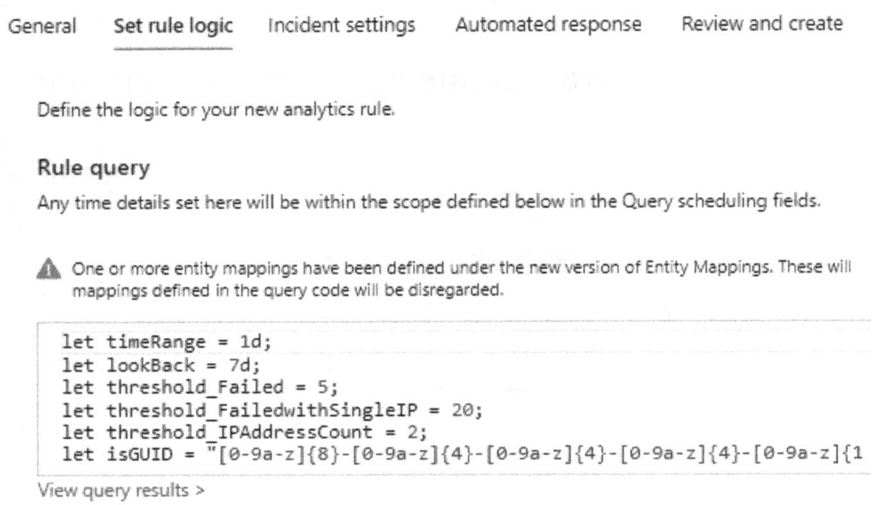

General   **Set rule logic**   Incident settings   Automated response   Review and create

Define the logic for your new analytics rule.

Rule query

Any time details set here will be within the scope defined below in the Query scheduling fields.

> ⚠ One or more entity mappings have been defined under the new version of Entity Mappings. These will
> mappings defined in the query code will be disregarded.

```
let timeRange = 1d;
let lookBack = 7d;
let threshold_Failed = 5;
let threshold_FailedwithSingleIP = 20;
let threshold_IPAddressCount = 2;
let isGUID = "[0-9a-z]{8}-[0-9a-z]{4}-[0-9a-z]{4}-[0-9a-z]{4}-[0-9a-z]{1
```

View query results >

Figure 9.50 – Analytics rule logic

6.   Next, click **Next : Incident settings** > and **Next : Automated response** >.

7.   On the **Automated response** tab, click **Add new** in the **Automation rules** section.

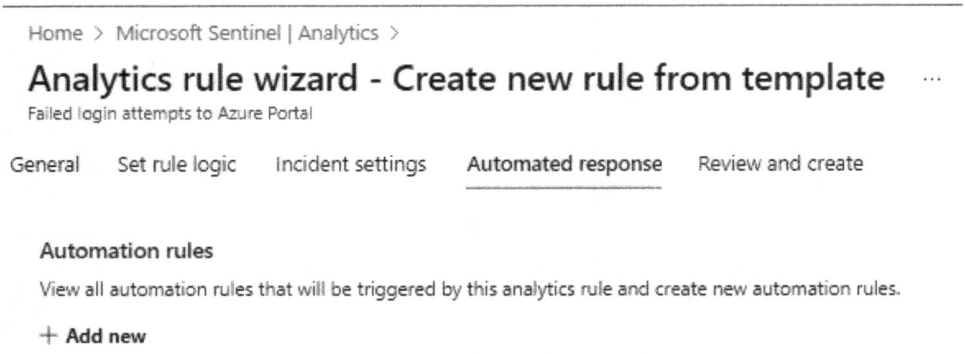

Home > Microsoft Sentinel | Analytics >

Analytics rule wizard - Create new rule from template   ⋯

Failed login attempts to Azure Portal

General   Set rule logic   Incident settings   **Automated response**   Review and create

Automation rules

View all automation rules that will be triggered by this analytics rule and create new automation rules.

+ **Add new**

Figure 9.51 – Analytics automated response

8.   Enter the **Automation rule name** value as required on the **Create new automation rule** blade, then set the **Actions** option to **Run playbook**.

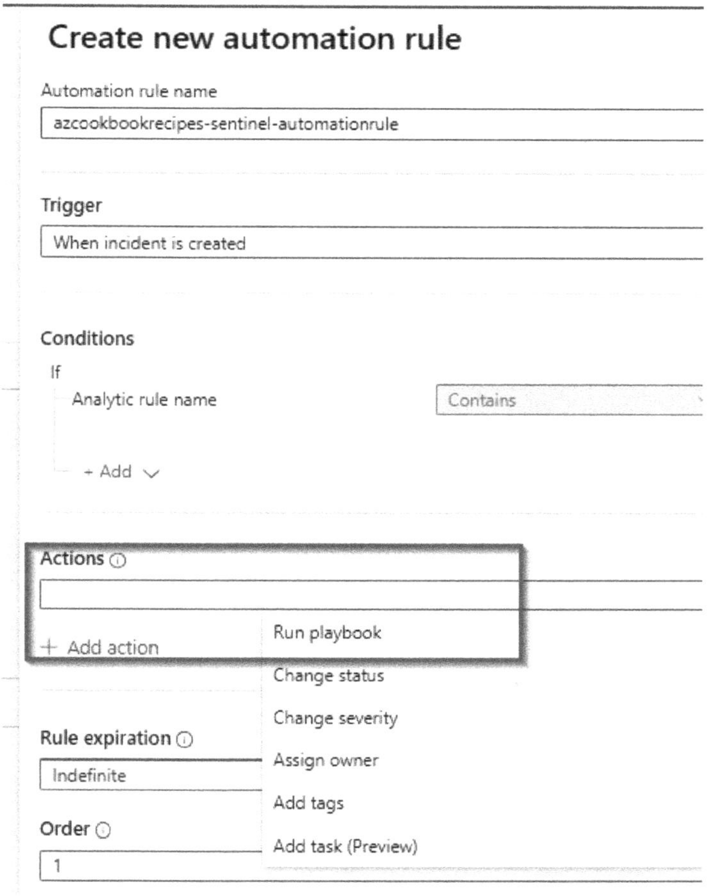

Figure 9.52 – Automation rule actions

9. Select the playbook we created in the previous recipe of this chapter.

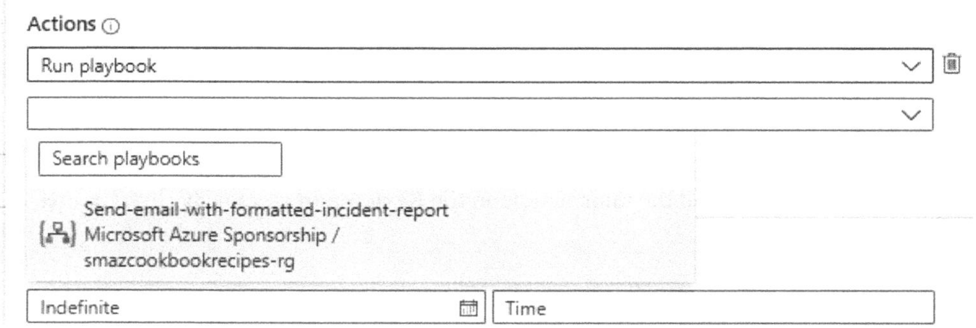

Figure 9.53 – Select the playbook

10. Then click **Apply**.

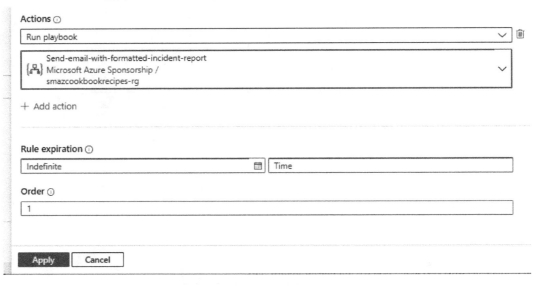

Figure 9.54 – Apply automation rule

11. You will now see the automation rule listed in the **Automated response** tab.

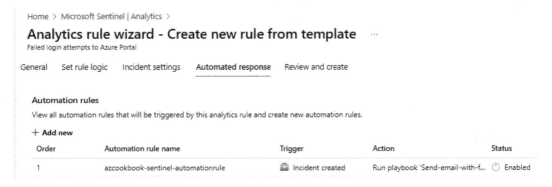

Figure 9.55 – Automation rule added

12. Click **Next : Review >**.

13. Once you have reviewed the information on the **Review and create** tab, click **Create**.

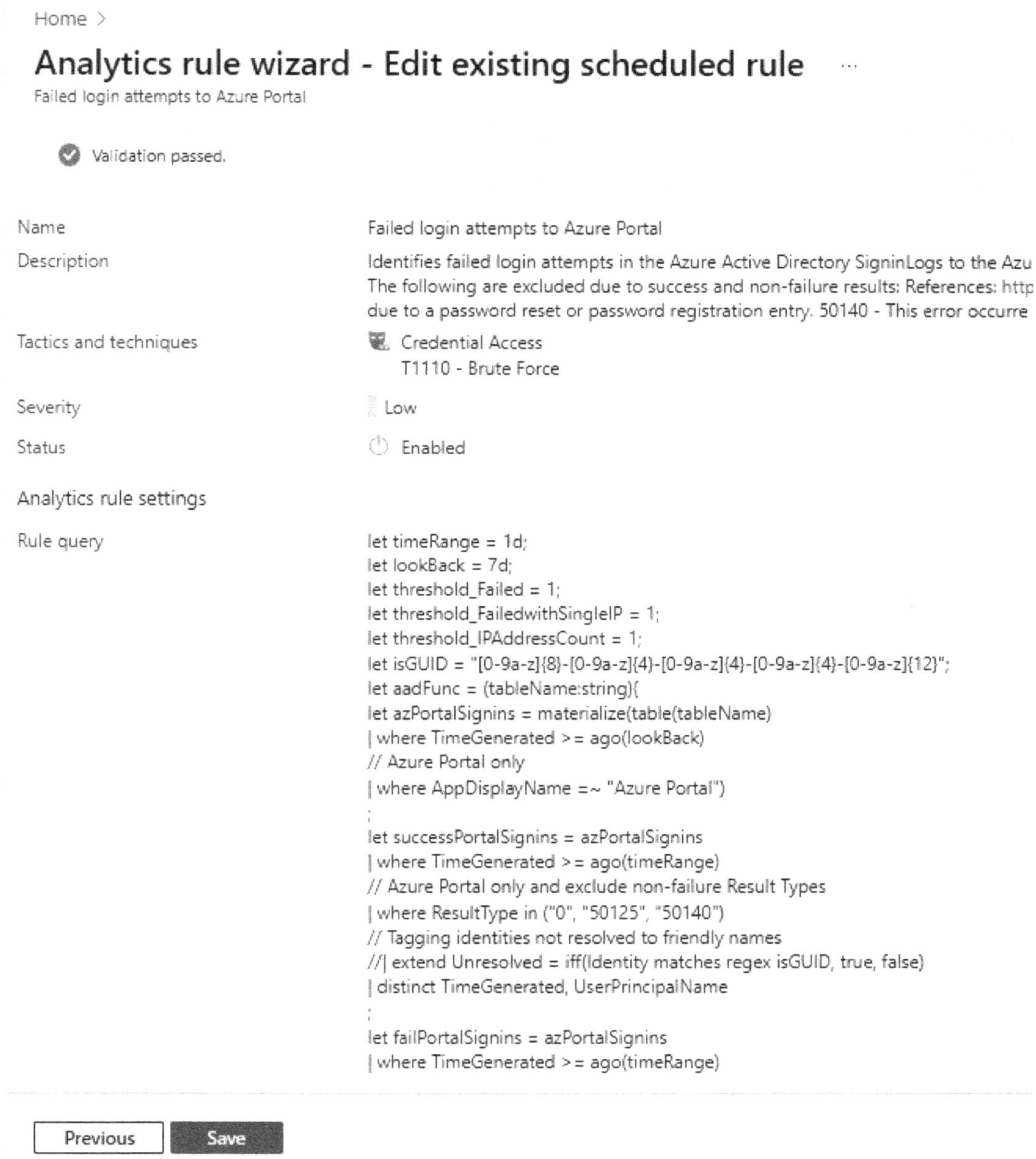

Home >

Analytics rule wizard - Edit existing scheduled rule   ...

Failed login attempts to Azure Portal

✓ Validation passed.

Name	Failed login attempts to Azure Portal
Description	Identifies failed login attempts in the Azure Active Directory SigninLogs to the Azu
The following are excluded due to success and non-failure results: References: http
due to a password reset or password registration entry. 50140 - This error occurre
Tactics and techniques	🐛 Credential Access
T1110 - Brute Force
Severity	Low
Status	⏻ Enabled

Analytics rule settings

Rule query

```
let timeRange = 1d;
let lookBack = 7d;
let threshold_Failed = 1;
let threshold_FailedwithSingleIP = 1;
let threshold_IPAddressCount = 1;
let isGUID = "[0-9a-z]{8}-[0-9a-z]{4}-[0-9a-z]{4}-[0-9a-z]{4}-[0-9a-z]{12}";
let aadFunc = (tableName:string){
let azPortalSignins = materialize(table(tableName)
| where TimeGenerated >= ago(lookBack)
// Azure Portal only
| where AppDisplayName =~ "Azure Portal")
;
let successPortalSignins = azPortalSignins
| where TimeGenerated >= ago(timeRange)
// Azure Portal only and exclude non-failure Result Types
| where ResultType in ("0", "50125", "50140")
// Tagging identities not resolved to friendly names
//| extend Unresolved = iff(Identity matches regex isGUID, true, false)
| distinct TimeGenerated, UserPrincipalName
;
let failPortalSignins = azPortalSignins
| where TimeGenerated >= ago(timeRange)
```

[Previous]   [Save]

Figure 9.56 – Analytics rule creation validation

14. You will be notified that the analytics rule was saved.

Figure 9.57 – Notification of success

15. On the **Analytics** screen, in the **Configuration** section, we can see the **Failed login attempts to Azure portal** rule in the **Active rules** section.

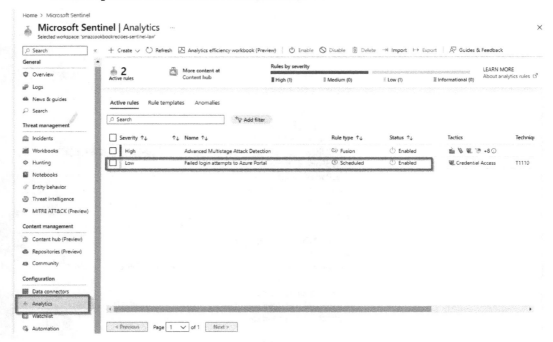

Figure 9.58 – Analytics active rules

16. After a short period of time, make some failed attempts to log in to the Azure portal to trigger the analytics rule logic.

Figure 9.59 – Failed portal login attempts

17. An incident will be created when the failed login attempts match the configured rule logic.

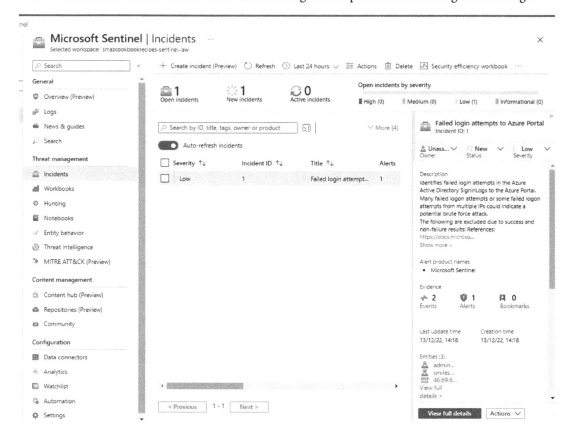

Figure 9.60 – Incidents

18. The playbook was also triggered, and an email notification was received.

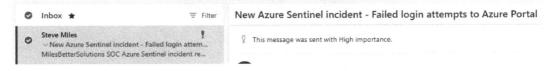

Figure 9.61 – Incident automated response notification

19. When the email notification is opened, further details of the incident can be found, such as the accounts that failed the Azure portal login process, an IP address, and a time stamp.

Figure 9.62 – Response notification email body

This task is completed.

How it works...

We saw how to add a data connector to collect signal data for this recipe. We then configured an analytics rule and viewed an incident triggered by the analytics rule logic.

See also

Should you wish to learn more about this and related topics, you can refer to the following Microsoft Learn articles:

- *Microsoft Sentinel documentation*: `https://learn.microsoft.com/en-us/azure/sentinel/`

- *Microsoft Learn training modules and paths*: `https://learn.microsoft.com/en-us/training/browse/?expanded=azure&products=azure-sentinel`

- *Microsoft Sentinel data connectors*: `https://learn.microsoft.com/en-gb/azure/sentinel/connect-data-sources`

- *Data collection best practices*: `https://learn.microsoft.com/en-gb/azure/sentinel/best-practices-data`

- *Detect threats out-of-the-box*: `https://learn.microsoft.com/en-gb/azure/sentinel/detect-threats-built-in`

10
Using Traffic Analytics

In the previous chapter, we covered recipes for effectively using Microsoft Sentinel.

In this chapter, you will learn how to collect NSG flow logs from **virtual machines** (**VMs**) to monitor and analyze network traffic.

By the end of this chapter, you will have learned the skills required to carry out the following recipe in secure Azure AD:

- Implementing traffic analytics

Technical requirements

This chapter assumes that you have an **Azure AD tenancy** and an **Azure subscription** from completing the recipes in previous chapters of this cookbook. If you skipped straight to this section, the information needed to create a new Azure AD tenancy and an Azure subscription for these recipes is included in the following list of requirements.

For this chapter, the following are required:

- A device with a browser, such as *Edge* or *Chrome*, to access the **Azure portal**: https://portal.azure.com

- An Azure AD tenancy and Azure subscription; you may use existing or sign up for free: https://azure.microsoft.com/en-us/free

- An **Owner role** for the **Azure subscription**

Terminology reference

We will start with some of the terminology used in **traffic analytics**:

- **Network security group** (**NSG**): This controls network traffic flow into and out of a VM via a network interface

- **NSG flow logs**: We can capture information about every packet that flows into and out of the VM (*ingress* and *egress*)

- **Log Analytics**: This service allows us to perform analytics on data sent to Azure Monitor and stored in a Log Analytics workspace

- **Network Watcher**: This is a network health service that allows us to monitor, view metrics, and diagnose network-level traffic

Now that we have covered some related terminology, we will move on to our first recipe for this section.

Implementing traffic analytics

Traffic analytics provides rich visual representations of *network packet information* made available by NSG flow logs. The NSG flow logs capture network traffic information, such as IP address *source* and *destination*, and the packet's port and protocol used.

This recipe will teach you how to implement traffic analytics with NSG flow logs.

Getting ready

This recipe requires the following to be in place:

- A device with a browser, such as Edge or Chrome, to access the Azure portal: `https://portal.azure.com`

- Access to an Azure subscription, where you have access to the Owner role

- A *Windows Server* **Azure VM** with an NSG to use with this recipe; we will step through creating this VM and NSG as a getting-ready task

Continue with the following getting-ready tasks for this recipe:

- Creating a VM

- Creating a Log Analytics workspace

A getting-ready task – creating a VM

Perform the following steps:

1. In the search bar in the Azure portal, type `virtual machines` and select **Virtual machines** from the listed **Services** results.

2. Click **Create** from the top-left menu bar on the **Virtual machine** screen and select **Azure virtual machine**.

3. On the **Basics** tab, under the **Project details** section, set the **Subscription** as required.

4. Click **Create new** for **Resource group**.

5. Enter a **Name** and click **OK**.

6. Under **Instance details**, set the following:

    - **Virtual machine name**: Type a name

    - **Region**: Select a region

    - **Availability options**: Select **No infrastructure redundancy required**

    - **Security type**: Select **Standard**

    - **Image**: Select **Windows Server 2019 Datacenter – X64 Gen2**

    - **Size**: Leave the default (or set it as required to reduce recipe costs.

7. Under **Administrator account**, set **Username** and **Password** as required.

8. Under **Inbound port rules**, set **Public inbound ports** to **Allow selected ports**.

9. Set **Select inbound ports** to **HTTP** (80), **HTTPS** (443), and **RDP** (3389).

10. Click **Next : Disks**, leave the default values, then click **Next : Networking**.

11. Under **Network interface**, leave the default values for **Virtual Network**, **Subnet**, and **Public IP**.

12. Ensure **NIC network security group** is set to **Basic**.

13. Leave **Select inbound ports** to the settings set in *step 9*: **HTTP** (80), **HTTPS** (443), and **RDP** (3389).

14. Tick the **Delete public IP and NIC when VM is deleted** box.

15. Click **Review + create**.

16. Click **Create** on the **Review + create** tab once validation has passed.

17. A notification will display that the resource deployment succeeded

The first getting-ready task for this recipe is complete.

Getting-ready task – creating a Log Analytics workspace

Perform the following steps:

1. In the search bar in the Azure portal, type `log analytics workspaces` and select **Log Analytics workspaces** from the listed **Services** results.

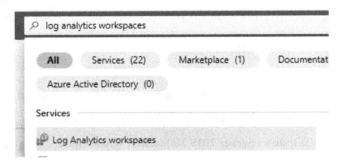

Figure 10.1 – Search for the resource

2. On the **Log Analytics workspaces** screen, click **Create** from the top-left menu bar.

3. On the **Basics** tab of the **Create Log Analytics workspace** screen, select a value in the **Subscription** and **Resource group** fields as required in the **Project details** section.

Home > Log Analytics workspaces >

Create Log Analytics workspace   ...

Basics   Tags   Review + Create

> ⓘ A Log Analytics workspace is the basic management unit of Azure Monitor Logs. There are specific considerations you should take when creating a new Log Analytics workspace. Learn more

With Azure Monitor Logs you can easily store, retain, and query data collected from your monitored resources in Azure and other environments for valuable insights. A Log Analytics workspace is the logical storage unit where your log data is collected and stored.

Project details

Select the subscription to manage deployed resources and costs. Use resource groups like folders to organize and manage all your resources.

Subscription * ⓘ	Microsoft Azure Sponsorship (8de2e9e8-de94-4feb-8a95-35b48b59... ⌄
└─ Resource group * ⓘ	azcookbookrecipes-rg ⌄
	Create new

Figure 10.2 – Create Log Analytics workspace

4. In the **Instance details** section, enter a value for **Name**, select a value in the **Region** drop-down menu, and then click **Review + Create**.

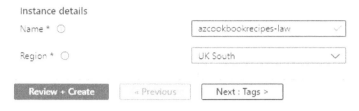

Figure 10.3 – Set Instance details

5. On the **Review + Create** tab, click **Create**.

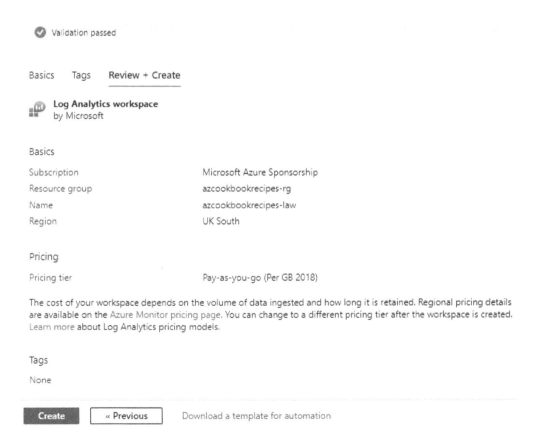

Figure 10.4 – Create a workspace

6. You will be notified that the deployment was successful.

The getting ready task for this recipe is complete.

You are now ready to continue the main tasks for this recipe of implementing Traffic Analytics.

How to do it...

This task consists of the following step:

- Implementing Traffic Analytics

Task – implementing Traffic Analytics

Perform the following steps:

1. Sign in to the Azure portal: `https://portal.azure.com`.
2. In the search bar, type `network watcher`; click **Network Watcher** from the list of services shown.

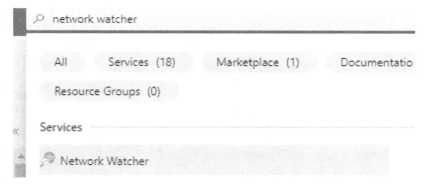

Figure 10.5 – Search for the resource

3. On the **Network Watcher** page, click **NSG flow logs** in the **Logs** section of the left-hand menu.

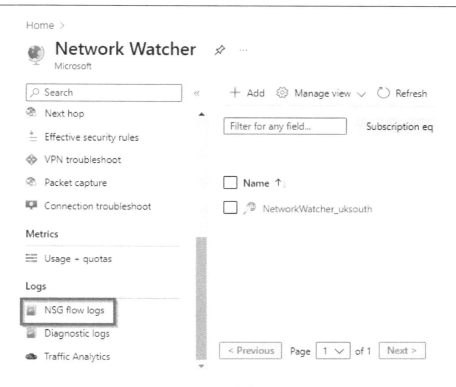

Figure 10.6 – NSG flow logs

4. From the **NSG flow logs** page, click **Create**.

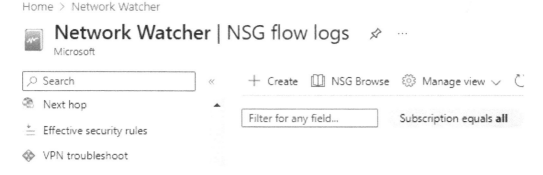

Figure 10.7 – Create NSG flow logs

5. Select a value in the **Subscription** drop-down menu as required from the **Basics** tab on the **Create a flow log** page, and then click **Select NSG**.

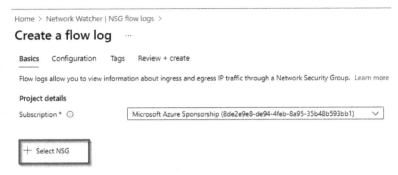

Figure 10.8 – Select NSG

6. Select the NSG created with the VM in the getting-ready task, then click **Confirm selection**.

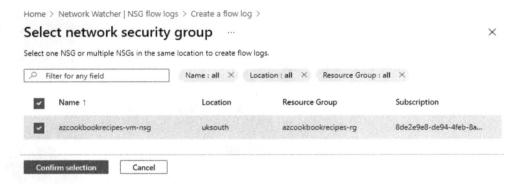

Figure 10.9 – Confirm selection of NSG

7. In the **Instance details** section, click **Create a new storage account**.

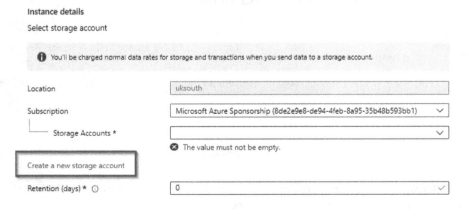

Figure 10.10 – Instance details settings

8. On the **Create storage account** blade, enter a value in the **Name** field, select a value from the **Resource group** drop-down menu, and then click **OK**.

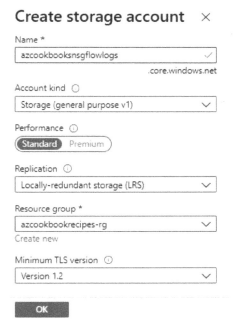

Figure 10.11 – Create storage account

9. Click **Next : Configuration**.

10. On the **Configuration** tab, in the **Traffic Analytics** section, check the **Enable Traffic Analytics** box. Set the **Traffic Analytics processing interval** to **Every 10 mins**; select the name of the Log Analytics workspace we created in the getting-ready task if not already selected in the **Log Analytics Workspace** drop-down menu, and then click **Review + create**.

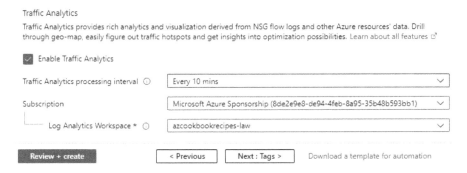

Figure 10.12 – Review and create a storage account

11. On the **Review + Create** tab, click **Create**.

12. You will be notified that the deployment was successful.

13. When you navigate back to the **Traffic Analytics** page, you will notice that you have to wait for some time for data to be logged.

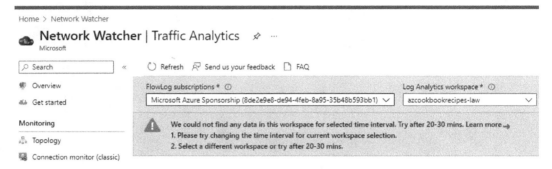

Figure 10.13 – Traffic Analytics blade

This task is completed. In the next task, we will clean up the resources created in this recipe.

Task – clean up resources

Perform the following steps:

1. In the search bar in the Azure portal, type `resource groups`, and select **Resource Groups** from the listed **Services** results.

2. On the **Resource groups** page, select the resource group we created for this recipe, and click **Delete resource group**; this will delete all the resources created as part of this recipe.

Figure 10.14 – Delete resource group

This task to clean up the resources created in this recipe is complete.

How it works...

For this recipe, we looked at implementing Traffic Analytics. The raw Network Watcher NSG flow logs are aggregated and stored in a Log Analytics workspace. These reduced stored logs then have *geography*, *security*, and *topology* enhancements added and then passed for analysis, allowing visualization of traffic patterns.

To perform NSG traffic analysis, you must have a Network Watcher enabled in each region where you have NSGs. We created a VM as a *getting ready task* that enabled a Network Watcher in our environment; an existing Network Watcher could be used if one already existed.

The following are usage scenarios and insights that can be gained with Traffic Analytics implemented:

- Find traffic hotspots
- Visualize traffic distribution by geography
- Visualize traffic distribution by virtual networks
- View ports and VMs receiving traffic from the internet

See also

Should you wish to learn more about this and related topics, you can refer to the following Microsoft Learn articles:

- *Traffic analytics*: https://learn.microsoft.com/en-us/azure/network-watcher/traffic-analytics
- *Usage scenarios*: https://learn.microsoft.com/en-us/azure/network-watcher/usage-scenarios-traffic-analytics
- *Traffic Analytics – frequently asked questions*: https://learn.microsoft.com/en-us/azure/network-watcher/traffic-analytics-faq
- *Azure Network Watcher documentation*: https://learn.microsoft.com/en-us/azure/network-watcher/
- *Introduction to flow logging for network security groups*: https://learn.microsoft.com/en-us/azure/network-watcher/network-watcher-nsg-flow-logging-overview

Index

www.packtpub.com

Subscribe to our online digital library for full access to over 7,000 books and videos, as well as industry leading tools to help you plan your personal development and advance your career. For more information, please visit our website.

Why subscribe?

- Spend less time learning and more time coding with practical eBooks and Videos from over 4,000 industry professionals

- Improve your learning with Skill Plans built especially for you

- Get a free eBook or video every month

- Fully searchable for easy access to vital information

- Copy and paste, print, and bookmark content

Did you know that Packt offers eBook versions of every book published, with PDF and ePub files available? You can upgrade to the eBook version at packt.com and as a print book customer, you are entitled to a discount on the eBook copy. Get in touch with us at customercare@packtpub.com for more details.

At www.packt.com, you can also read a collection of free technical articles, sign up for a range of free newsletters, and receive exclusive discounts and offers on Packt books and eBooks.

Other Books You May Enjoy

If you enjoyed this book, you may be interested in these other books by Packt:

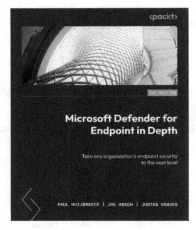

Microsoft Defender for Endpoint in Depth

Paul Huijbregts, Joe Anich, Justen Graves

ISBN: 9781804615461

- Understand the backstory of Microsoft Defender for Endpoint
- Discover different features, their applicability, and caveats
- Prepare and plan a rollout within an organization
- Explore tools and methods to successfully operationalize the product
- Implement continuous operations and improvement to your security posture
- Get to grips with the day-to-day of SecOps teams operating the product
- Deal with common issues using various techniques and tools
- Uncover commonly used commands, tips, and tricks

Cloud Security Handbook

Eyal Estrin

ISBN: 9781800569195

- Secure compute, storage, and networking services in the cloud
- Get to grips with identity management in the cloud
- Audit and monitor cloud services from a security point of view
- Identify common threats and implement encryption solutions in cloud services
- Maintain security and compliance in the cloud
- Implement security in hybrid and multi-cloud environments
- Design and maintain security in a large-scale cloud environment

Packt is searching for authors like you

If you're interested in becoming an author for Packt, please visit `authors.packtpub.com` and apply today. We have worked with thousands of developers and tech professionals, just like you, to help them share their insight with the global tech community. You can make a general application, apply for a specific hot topic that we are recruiting an author for, or submit your own idea.

Share your thoughts

Now you've finished *Azure Security Cookbook*, we'd love to hear your thoughts! Scan the QR code below to go straight to the Amazon review page for this book and share your feedback or leave a review on the site that you purchased it from.

`https://packt.link/r/1804617962`

Your review is important to us and the tech community and will help us make sure we're delivering excellent quality content.

Download a free PDF copy of this book

Thanks for purchasing this book!

Do you like to read on the go but are unable to carry your print books everywhere?

Is your eBook purchase not compatible with the device of your choice?

Don't worry, now with every Packt book you get a DRM-free PDF version of that book at no cost.

Read anywhere, any place, on any device. Search, copy, and paste code from your favorite technical books directly into your application.

The perks don't stop there, you can get exclusive access to discounts, newsletters, and great free content in your inbox daily

Follow these simple steps to get the benefits:

1. Scan the QR code or visit the link below

https://packt.link/free-ebook/9781804617960

2. Submit your proof of purchase
3. That's it! We'll send your free PDF and other benefits to your email directly

www.ingramcontent.com/pod-product-compliance
Lightning Source LLC
Chambersburg PA
CBHW062050050326
40690CB00016B/3038